Gender, Sexual Citizenship and Epistemic Injustice in the Caribbean

Charmaine Crawford

Gender, Sexual Citizenship and Epistemic Injustice in the Caribbean

palgrave
macmillan

Charmaine Crawford
Department of Africana Studies
Kent State University
Kent, OH, USA

ISBN 978-3-031-83492-9   ISBN 978-3-031-83493-6   (eBook)
https://doi.org/10.1007/978-3-031-83493-6

This Palgrave Pivot imprint is published by the registered company Springer Nature Switzerland AG.
The registered company address is: Gewerbestrasse 11, 6330 Cham, Switzerland

If disposing of this product, please recycle the paper.

Dedicated to the loving memory of A.C. and C.D.
And to
all the Caribbean LGBTQ activists in the region who
dared to transgress

ACKNOWLEDGEMENTS

There is no such thing as a single-issue struggle, because we do not live single-issue lives. –Audre Lorde.

Writing this book has been both a personal and academic journey for me. I lived and worked in Barbados for over 10 years, so I had many discussions and encounters with individuals who inspired me to embark on this book project. As a black lesbian feminist of Caribbean descent, my transnational connections span the Caribbean, Canada and the United States. The disruption of colonial and post-colonial heteropatriarchal constructions of sexual citizenship currently in the region is evident by the vibrant intersectional politics going on with a new generation of feminist and lesbian, gay, bisexual, transgender and queer (LGBTQ) activists fighting for gender and sexual equality in their societies. Therefore, in this publication I attempt to trouble hegemonic notions of sexual citizenship through a decolonial queer feminist episteme.

I would like to thank the LGBTQ groups and leaders for their tireless work in challenging homophobia and transphobia in the region. I would like to extend appreciation to the co-founders of Barbados—Gays, Lesbians and All-Sexuals against Discrimination (B-GLAD), Donnya Piggott and Ro-Ann Mohammed, for standing up for justice in the face of ridicule and for their meaningful interactions with me. I would also like to extend my thanks to Kenita Placide, founder and executive director of Eastern Caribbean Alliance for Diversity and Equality (ECADE) and former Executive Director of United and Strong (St Lucia), for their leadership on the regional and international level in raising the visibility of LGBTQ rights, especially for lesbians, transgender women, and gender non-binary persons, and for promoting youth leadership and mentorship. I would like to acknowledge the activism of the late Colin Robinson, founder of the Coalition Advocating for the Inclusion of Sexual Orientation (CAISO), who fought for legislative and cultural change to tackle bigotry and discrimination in Trinidad and Tobago while holding onto a deep love for his country.

My acknowledgement to my former colleagues at the University of the West Indies is manifold because their work has been valuable to my own academic explorations. I want to extend special thanks to trailblazing Caribbean feminists and gender scholars Peggy Antrobus, Eudine Barriteau, Rhoda Reddock, Verene Shepherd, Patricia Mohammed and Hilary Beckles, for their mentorship, whose scholarship and activism in the areas of women rights, women/gender and history gender and development, and feminist theorising and praxis in the region are unmatched. As knowledge producers, your contributions to academia, public policy and civil society have helped in further democratising Caribbean nations and in educating a new generation of critical thinkers. I would also like to thank the staff and faculty (Veronica Jones, Halimah DeShong, Tonya Haynes, Leigh-Ann Worrell and Daniele Bobb) of the Institute of Gender and Development Studies: Nita Barrow Unit (IGDS: NBU) at the University of the West Indies Cave Hill Campus who have championed teaching, research and outreach in the areas of women, gender, and sexual and feminist studies in the Commonwealth Caribbean and beyond. My time spent working at the Institute contributed to my personal, professional and academic development and it gave me the opportunity to be involved in multiple outreach activities around gender justice, women's rights and LGBTQ rights that heightened my awareness across social movements. I extend special thanks to Shari-Inniss Grant, a former research assistant at the Institute, for collaborating on the Sexual Equality and Human Rights Youth project that documents the experiences of LGBTQ youth in Barbados that was initially sponsored by the British High Commission in Barbados, which is captured in some parts of this book. Finally, I extend my appreciation to the staff and faculty at the other IGDS units in Jamaica and Trinidad and Tobago.

I would like to extend my sincerest thanks to M. Jacqui Alexander for addressing and complicating the relationship between coloniality, gender, sexuality and state power within local and global hegemonies. You have been a source of strength and inspiration to me both academically and spiritually. Your critical analysis of sexual citizenship and the few conservations that we had over the years helped me delve deeper into exploring the relationship between sexual citizenship, LGBTQ rights and epistemic justice. In addition to this, scholarship on gender, sex and citizenship by feminist legal scholar Tracy Robinson, as well as the strategic litigation of the University of the West Indies, Rights Advocacy Project (U-RAP), must be lauded for exposing and overturning discriminatory laws against women and LGBTQ and transgender persons. Thank you for your common-sense approach of critiquing laws which makes them accessible to those of us who are not legal experts. I also want to extend thanks to Latoya Lazarus (SALISES) for her scholarship in problematising heteronormativity and Christian citizenship in the Caribbean and for her input on new ideas that I was exploring on sexual citizenship and LBT women for another piece.

Furthermore, I want to extend my deepest gratitude to my former PhD supervisors, Kamala Kempadoo and David Murray, at York University, Toronto, Canada for their guidance and support throughout the years. I would like to thank Kamala for your cutting-edge research on sex work, sex tourism and sexuality studies in the Caribbean which has helped me think outside the box with my own ideas, and thanks to David for his investigations on homosexuality and transgenderism in Barbados as well as human rights and LGBTQ activism. Moreover, I would like to extend my sincerest thanks to the Department of Africana Studies and the College of Arts and Sciences at Kent State University for its support and resources in completing this book project.

Finally, I want to extend a heartfelt thank you to my family (Noel, Sonia and Sonelle) and forever friends (Tennisha Morris, Andrea Jones and Amoaba Gooden) for their support and love through this arduous journey that seemed never ending. Across different borders and locations, you inspired me to preserve and to exceed my own expectations for this book.

CONTENTS

1 Introduction   1

2 Sexual Citizenship and the Politics of Exclusion   19

3 Epistemic Injustice, Sexual Inequality and LGBTQ
Realities   39

4 The Precarity of Sexual Citizenship: Hermeneutical
Injustice, the Law and LGBTQ Rights   61

5 "It's a Girl Thing": Problematising Female Sexuality,
Gender and Lesbophobia in Caribbean Culture   105

6 Decolonising and Queering Caribbean Families   129

Index   145

CHAPTER 1

Introduction

Abstract In this chapter, I outline the conceptual and theoretical frameworks as well as the methodological approach used in this study to examine gender, sexual citizenship and epistemic injustice as it relates to LGBTQ experience in the Commonwealth Caribbean. Decolonial feminist queer perspectives as well as social justice hermeneutics encapsulate this multifarious interdisciplinary research project. In doing so, I deconstruct a Marshallian citizenship model rooted in Western liberalism that presupposes a linear path to modern political development and civic engagement that deceptively constructs a universal citizen who is bearer of a multitude of rights and whose identity is neutrally configured within society. This not only obscures the gender, race and class hierarchies and their intersections in shaping which persons belong or do not belong to the nation, but it also conceals how intimate life and sexual citizen are integral to how the state maintains its power by organising and regulating socio-sexual relations, family and reproduction. In addition to this, the connection between colonial/post-colonial systems of power, interlocking oppressions and epistemic injustices is made clear in this chapter. Heteropatriarchal tropes relegate Caribbean LGBTQ persons to the margins of sexual citizenship, in turn normalising stigma and discrimination against them. Consequently, this produces epistemic injustices that discredit them as credible knowers of their own realities as well as devaluing their worth in society, legitimising their outsider status as sexual citizens. The counter to this, the epistemic resistance launched by LGBTQ and feminist activists for sexual rights, is a testament to the vibrancy of intersectional politics as well as continuity of social justice activism in the region.

Keywords Audre Lorde and breaking silences • Hegemonic knowledge and deconstructing citizenship (T.H. Marshall model) • Sexual citizenship • Black/Caribbean feminisms • Anti-colonialism • Intersectionality • Decolonial feminist queer theorising • The law, epistemic injustice and Caribbean LGBTQ rights

> What are the words you do not have yet? What do you need to say? What are the tyrannies you swallow day by day and attempt to make your own, until you will sicken and die of them, still in silence? Perhaps for some of you here today, I am the face of one of your fears. Because I am woman, because I am Black, because I am lesbian, because I myself—a Black woman warrior poet doing my work—come to ask you, are you doing yours. (Lorde, 2017a, 2017b, p. 3)

There is a kind of existential crisis that individuals on the margins experience when their existence seems to be marred by oppressive forces that seek to subordinate or hold them down because of who they are. As a black lesbian feminist of Caribbean descent, I am drawn to Lorde's words about language, silence and transformation. If authentic power is harnessed from within, as Audre Lorde so eloquently described in *Uses of the Erotic*, then there is something powerfully redemptive about using one's *voice* to tap into one's authentic self and to express our innermost feelings, desires and frustrations. In her piece "The Transformation of Silence into Language and Action," Lorde instructs us to break free of silences brought on by fear, domination and violence, which she feels ultimately immobilise us. She states, "I have come to believe over and over again that what is most important to me must be spoken, made verbal and shared, even at the risk of having it bruised or misunderstood" (Lorde, 2017a, 2017b, p. 1). In dealing with her own mortality after her breast cancer diagnosis, Lorde laments about her fears and regrets in not speaking up when it mattered most in her life. She candidly encourages individuals to give voice to their pain and injustices before they get swallowed up by rage or face the risk of others speaking for them or about them to their own detriment.

Lorde writes, "I was going to die, if not sooner than later, whether or not I had ever spoken myself. My silences had not protected me. Your silence will not protect you" (ibid., 2017, p. 2). Knowledge is power is a popular sentiment, but Lorde's account considers how the *word* or *language* (oral, written and symbolic) is just as powerful in not only crafting, assembling and articulating knowledge but in transforming the personal into political action against oppression. The consciousness-raising efforts of feminist, black nationalist and anti-colonial struggles during the twentieth century were crucial in providing spaces and tools for women and black, indigenous and people of colour to engage in self-discovery in naming and voicing their concerns, which then gave them an opportunity to participate in collective action to challenge inequality.

While Lorde recognises that subjugated people have legitimate reasons for not always sharing their knowledge due to the fear of reprisals or the heightened vulnerability that comes with visibility (being heard and seen), she nonetheless sees silence as a more dangerous alternative because it signals the death of the authentic self. As a black lesbian feminist, Lorde gives voice for her love and sexual desire for women by invoking the word *Zami*, which is a Carriacou term that describes women who come to love each other through personal interactions. In challenging monolithic construction of identities, Lorde spoke out about the importance of celebrating each other differences, based on the multiplicity and intersectionality of identities, but she knew that was not welcomed by everyone. She felt less valued and trusted by others who held on to some facet of their privilege (whether gender, race, class sexuality and/or ability) within white patriarchal structures in the United States or in social movements (Western feminism and black nationalism) where a race/gender divide existed.

Lorde's perspective on silence, language and social transformation is valuable because it provides an opening to discuss to how LGBTQ[1] in the

---

[1] Given the diversity of cultural diasporas in the Caribbean, and specifically the black Atlantic experience with its diasporic and transnational connections in European and North America, there is a variation of gender and sexual identities and practices both named and unnamed that individuals attribute to themselves in the region. Same-sex and gender nonconforming identities that are named may be derived from the local vernacular, while other sources of naming include Western descriptors or terminology that characterise non-binary and non-heteronormative gender and sexual identities and expressions that deviate from the status quo. This is the reality of post-colonial queer Caribbean identities. This book incorporates popular terminology used by LGBTQ persons to define themselves in the region regardless of the origin of the nomenclature.

Commonwealth Caribbean have to deal with the paradox by strategically using silence to protect themselves from harm while at the same time optimising the use of their voice when they have an opportunity to do so to validate their experiences and to politicise their grievances against being treated as second-class citizen in their countries. In *Citizenship from Below: Erotic Agency and Caribbean Freedom*, Sheller states that "race, ethnicity, gender and sexuality are bodily practices of differentiation that surface at the intersections of multiple forms of state ordering, moral regulation, self-discipline, and systems of governance that endorse and make possible regimes of free citizenship" (2012, p. 22). Accordingly, citizenship, and by extension sexual citizenship, in the Caribbean, and elsewhere in the Americas, cannot be separated from systems of oppression such as the legacy of European conquest, colonialism and slavery that consolidated Western hegemonic power in the region.

The restrictions placed on black people, women and LGBTQ persons, and the intersection of these identities, such as black queers, as sexual citizens have been historically institutionalised through racialised, gendered and heteronormative ideologies that are not only present in the public domain but are also very present in private and intimate spaces. Sexual rights include the right to the recognition of one's sexual identity, a right to pleasurable sex and to choose one's sexual partner, freedom from sexual harm and coercion, and access to sexual health services to ensure bodily well-being and safety (Dixon-Mueller et al., 2009). The infringement of the sexual rights of LGBTQ persons in the Commonwealth Caribbean is objectionable because it has led to the policing and criminalisation of sexual conduct-based and other behaviours through and anti-buggery laws,[2] non-recognition of same-sex unions and families and no protection against discrimination on the basis of gender identity/expression and sexual orientation under state law. Full racial and cultural liberation from our colonial masters must include upending restrictive gender and sexual

---

[2] Buggery refers to sex per *anum*. The offence includes consensual and non-consensual anal sex between adults. Buggery deals with penetration by a male sexual organ so depending on the country the offence may include male to male sexual activities and male to female sexual activities (Robinson, 2009). To date, six Caribbean Commonwealth countries still retain buggery laws: Jamaica, Grenada, St Vincent and the Grenadines, Dominica, Guyana and St Lucia. The sentence for buggery can range from 1 to 25 years and Guyana is the only country that has life imprisonment for the offence. Other laws such as serious indecency criminalise other non-penetrative sex acts (such as oral sex and masturbation) which include consensual sexual relations between adult females.

norms that intersect with racism and ethnocentrism that seek to oppress certain groups of people. Alexander (2005) aptly reminds us that "antico-lonial nationalism had taught us well about heterosexual loyalty, a need so great that it reneged on its promise of self-determination, delivering crimi-nality instead of citizenship" (ibid., pp. 260–261). While separate sphere ideology creates an illusionary split between our public and private lives when it comes to intimacy, whereby the latter is presumed to be outside the reach of state interference, ironically, sexual citizenship has never been neutrally configured or existed separately from the state, law and political economy. In fact, the exploited and uncompensated labour of black women and queers has served as a basis of a colonial and post-colonial logic and project in establishing parameters of nationhood and citizenship. This is why critical social theory such as that of decolonial and anti-racist feminist perspectives in necessary in investigating and critiquing the rela-tionship between coloniality, race and gender and class within systems, institutions and social relations (Hill Collins, 2019; Mohanty & Carty, 2018; Sheller, 2012).

Purpose and Conceptual Framework

In this book, I interrogate sexual citizenship and LGBTQ rights in the Commonwealth Caribbean. I do so by employing a decolonial queer femi-nist perspective to problematise how race, gender, sexuality and class inter-play and are operationalised through matrices of power in shaping intimate life through the family, religion, media, state and law. I will also examine how LGBTQ persons are resisting prejudice and discrimination in their intimate lives when it comes to sexual conduct, socio-sexual unions and family. The contradictions of sexual citizenship in Caribbean modernity and other parts of the global South operate through dialectical relations of domination and subordination from colonial and post-colonial power struggles. Given this, I will critically analyse Caribbean LGBTQ persons as knowledge producers in generating new ideas (counter-knowledge) to challenge heteropatriarchal ideologies and structures as well as seeking to improve their lives and societies. As Lorde reminds us, silence turned into language (voice) and action can be liberating. Really, the fight for sexual agency and rights in the region is an iterative process, occurring through diverse subjugated racialised—gendered-sexed experiences that are episte-mologically valuable. Thus, the undoing of homophobia and transphobia requires a multifaceted interdisciplinary approach that not only addresses bigotry on an individual level but also seeks to overturn hegemonic knowledge and practices that are discursively produced, normalised and disseminated through different systems of power to obscure LGBTQ subjectivities and to discredit them as credible sexual citizens. By exploring the relationship between sexual citizenship, epistemic injustice and LGBTQ resistance in the Commonwealth Caribbean in the twenty-first century, there is an opportunity to imagine and create new ways of being and belonging in our intimate lives and societies that foster more fairness and inclusion.

Western Hegemony, T.H. Marshall and Citizenship for the Few

Both the formal and symbolic representations of citizenship reflect ideas about the national identity of a group of people based on shared commonalities, nationality, ethnicity and culture. Glenn notes that "community members participate in drawing the boundaries of citizenship and defining who is entitled to civil, political, and social rights by granting or withholding recognition" (2011, p. 3). Democratising citizenship, and by extension sexual citizenship, in the Commonwealth Caribbean has been a work in progress in the region because so much of the process relies on destabilising exclusionary notions of belonging and nationhood established from a colonial past which were integrated into post-independent Caribbean nation-states. Accordingly, to establish the context in discussing sexual citizenship and LGBTQ rights in the contemporary Caribbean, I situate its precarious development in broader political, socio-cultural and economic relations that shaped dominant citizenship rights in Western modernity.

In examining citizenship and social class in post-World War II Britain, sociologist T.H. Marshall notes that "citizenship requires a direct sense of community membership based on loyalty to a civilization that is a common possession. It is a loyalty of free men endowed with rights and protected by a common law" (2009, p. 151). He goes on to conceptualise citizenship as comprising three major rights: civil or legal rights, political rights and social rights. Firstly, Marshall reasons that the consolidation of civil (legal) rights arose in the eighteenth century and coincided with the Age of Enlightenment in Europe with liberalism and democratic principles at its core. Individuals (citizens) are conceived as rational beings who have the autonomy to choose what is best for them if their choices do not deny others the independence to do the same. Equal opportunity and rights are extended to persons to exercise their freedom of choice through the law and state. Civil rights are enshrined through the right to justice and liberty of the person, freedom of speech, assembly and religion, the right to own property and other concomitant entitlements. Secondly, Marshall notes that political rights emerged out of the nineteenth century and focused on issues around governance and state power. Political systems and participation were institutionalised through government bodies and the electoral process with the right to vote. Finally, Marshall links social rights emerging in the twentieth century with the rise of the welfare state as protector of the poor. Individuals were entitled to socio-economic security to allow their full participation in civil society through the state administering social benefits.

A major criticism of the Marshallian citizenship model is that it promotes a Western liberal perspective that presupposes a linear path to modern political development and civic engagement based on a functionalist approach to citizenship. If citizenship, as Marshall states, "is a loyalty of free men endowed with rights and protected by a common law" (2009, p. 151), then his model is limited to a few because it does not take into consideration the fact that not all individuals have the same rights, nor do they have the same resources to attain rights due to social inequality based on race, gender, sexuality, class, nationality and the like. Marshall's citizenship model deceptively constructs a universal citizen who is bearer of a multitude of rights and whose identity is neutrally configured within society. Wilson argues that "formal citizenship, within political theory, does not refer to recognition of belonging by society. It refers to recognition of belonging by the state. Articulating a desire, or making a claim, for citizenship does not make one a citizen" (2009, p. 75). Wilson's point reveals the exclusionary politics associated with claims to citizenship. Taking this point further on a global scale, Western nation-building, citizenship and political theory in the project of modernity are entangled with imperialism, colonialism, slavery and racism that resulted in the conquering of land and resources of indigenous peoples, as well as subjugating them to slavery and genocide in pursuit of gold and profit from the fifteenth century onward. These atrocities against indigenous peoples established the economic and racist rationale for dehumanisation and exploitation of African labour through the Transatlantic Slave Trade in the mass production and exportation of sugar and other raw materials from the colonies to the metropole. This mercantilist model drove industrial capitalist development in the West benefitting European (white), propertied men who consolidated wealth, power and rights as de facto citizens of their countries and territories over non-European/white populations.

If class (capital) and race were major factors in shaping hegemonic Western citizenship, then gender was significant in doing so too. In *The Sexual Contract*, Carol Pateman (1988) critiques social contract theory that emerged out of a Western moral and political philosophical tradition that relates to the personal autonomy with which individuals must make agreements with the state in exchange for rights and protection. This social contract forged out of the breastplate of modernity did not fully include women (read *white women*) in civil society because the *original contract* that was rooted in patriarchal prerogatives, or what she refers to as "men's freedom and women's subjection" (1988, p. 10), preceded it. Pateman dismisses the notion of universal freedom espoused in contract theory as fable or "political fiction" that is premised on unequal power relations of domination and subordination (ibid., p. 15). White middle-class women were politically and economically disenfranchised in their societies as second-class citizens. Gender and sexual oppression converged for white middle-class women because they lacked sexual autonomy as their sexuality and reproductive labour appropriated and exploited hetero-sexual partnerships within the institution of marriage. The rallying cry for gender equality led by white liberal feminists sparked the first and second wave women's movements for the right to vote and women's rights from the eighteenth to mid-twentieth century in Europe and North America. This challenge to white capitalist patriarchy unfolded as black and brown women were either kept outside or at the margins of these movements due to the racism and classism in white feminist organising.

Sexual Citizenship, Heteropatriarchy and Coloniality

While the Marshallian citizenship model of political and civil rights and responsibilities tends to magnify divisions between public and private life by deceptively presenting the state as a non-interventionist actor in the intimate lives of individuals, Glenn argues that "citizenship affects public life in such areas as political participation and development of state policy; it also affects private life, including family and interpersonal relations" (2011, p. 2). Sexuality permeates throughout all aspects of human relations, and it is co-constituted with gender, race and class in shaping our "erotic desires, practices and identities" and in elevating "aspects of personal and social life which have erotic significance" (Jackson & Scott, 1996, p. 2). Sexuality is an integral part of being human, and ultimately the state seeks to discipline and control differently configured gendered and sexed bodies and intimacies for its interest (Cooper, 1995). Although intimate life and citizenship are intertwined in a multitude of ways, feminists argue that this fact is often overlooked because of the personal and often private (hidden) dimensions of intimate life whereby women's sexuality and same-sex desire gets privatised under capitalism. Chateauvert states that "sexual citizenship means the adult right to organize one's sexual life and household as one desires, and to have one's privacy respected and recognized in law and social policy. Sexual citizenship means the right of mature adults to engage in consensual sexual relations without a marriage license (that is, the sanction of the state)" (2008, p. 199). While this general definition of sexual citizenship enshrines personal rights and privacy associated with consensual adult sexuality, such rights are not equally protected and applied to all.

Richardson states that sexual citizenship is difficult to pin down because "both 'citizenship' and 'sexual' are ambiguous terms whose meanings are contested" (2000, p. 86). While Evans utilises a political economy perspective in understanding sexual citizenship by looking at its material significance and the heightened commodification of sexuality under advanced capitalism, feminists and queer scholars[3] have

---

[3] Emerging from Western white academics in the late 1980s to early 1990s, queer theory is a postmodernist concept that refers to deconstructing fixed categories of gender and sexuality based on social binaries (man/women or heterosexual/homosexual). Queering as a process attempts to capture the ambiguities of gender, sexuality and erotic desires. Judith Butler (2011) notes that the body is not a definitive site of gender but instead it is an iterative site that creates gendered expressions though interactions and performances. In queer theory, it is more worthwhile to create meaning about our bodies, desires and sexual practices than it is to attach an identity or name to the sexual practice (gay or lesbian). Queer theory dislodges the notion that there is an essence to sex, gender and sexuality because these areas are not fixed, and they are constantly changing (Epstein, 2002). But black queer scholars have critiqued white queer scholars for blind spots on race and racism that do not take into consideration non-white LGBTQ persons and intersecting oppressions that they face in their communities and wider mainstream society.

While queer is a term derived from Western sexuality scholars, this does not mean that fluid and diverse gender and sexual identities and expressions did not exist in the Caribbean and elsewhere in the global South prior to the coining of the term (Wekker, 2006; Wieringa & Sivori, 2013). In this book I will use the term queer as well as local terms as descriptors for unconventional or non-conforming gender and sexual expressions, performances and non-binary identities.

problematised the ambiguities of citizenship and sexuality, as well as the tenuous relationship between the two concepts, in their theorising and collective action against patriarchy, heterosexism and homophobia. Richardson adds that sexual citizenship is entangled with the "institutionalization of heterosexuality" in public and private life (2000, p. 107). Patriarchy and heterosexism converge in ordering sexuality and reproduction through marriage and the heterosexual nuclear family that exploits women's sexuality and reproductive labour as wives and mothers and denigrates homosexuality and all non-procreative sex acts. Foucault (1978) discusses the normalisation of heterosexuality through a bourgeoise capitalist division of labour between the eighteenth and nineteenth centuries in the West. This was done via the demonisation of homosexuality as a perversion through the biopolitics of the state, medicine and religion. Moralistic indictments about homosexual perversions whether through religious puritanism or state were suppressed and silenced through disciplinary measures against individuals who did represent or conform to a heteronormative and bourgeoise capitalist division of labour and way of being that could be readily exploited or reproduced in maintaining the social order. Richardson (2000) notes that sexual rights activism gained traction in Western nations due to feminist activism around abortion and other reproductive rights and gay and lesbian activism around sexual conduct-based rights such as the decriminalisation of homosexuality during the 1960s.

From a subaltern perspective, Wieringa and Sivori state that through a Western gaze in studying sexuality in the global South, "non-Western sexualities acquired meaning strictly within bounds of indigenous cultures, reinforcing their exterior and interior status vis-à-vis the uncontested authority of modern Western knowledge" (2013, p. 5). So sexual citizenship for formerly colonised people from marginalised groups is an unfinished project because as outliers to the dominant order they simultaneously must pursue, resist and re-invent it to usurp its exclusionary tendencies that contribute to social inequalities and disenfranchisement. It is for this reason that feminists and critical race scholars have languished over the precarity of full citizenship for women due to gender inequality and the poor and formerly colonised black and brown peoples of the world due to racial and economic exploitation. During colonialism and slavery, black sexuality was pathologised as animalistic and deviant compared to white sexuality. Due to a racist colonial project in the Caribbean,

African-Caribbean men and women also had to deal with the demonisation of their sexuality as animalistic compared to white sexuality. Thus, an anti-colonial analysis is required in interrogating sexual citizenship. Thomas (2007) notes that Foucault's work on the history of sexuality should be seen as a critical analysis of sexuality in Western societies instead of being proscriptive for all of humanity based on a linear trajectory. He goes on to argue that "a chronology of empire dictates a normative genealogy of desire, indeed a colonial *telos* for individuals or human beings" (2007, p. 3). Thus, the exploration of sexual citizenship in this book uncovers the gaps in knowledge and analysis that have overlooked black queer embodiment and intimacies in the Caribbean.

DECOLONIAL FEMINISM, QUEER THEORISING AND PRAXIS

To explore the complexity of sexual citizenship and LGBTQ rights in the Commonwealth Caribbean, I will use a decolonial, queer feminist theoretical framework to interrogate the intersections of race, gender, sexuality and class that operate through matrices of power to regulate sexual relations and intimate life in the region. The multi- and interdisciplinary scope of this research project requires an eclectic theoretical framework that can bridge gaps in knowledge in studying social phenomena across different disciplines. In addition to this, since the hegemony of formal citizenship is intertwined with ethno-racial, heterosexual, masculinist prerogatives derived from colonialism and global capitalism, an intersectional analysis of sexual citizenship is necessary in identifying and challenging the interlocking oppressions that have restricted sexual freedoms of women and LGBTQ persons in the Commonwealth Caribbean. Decolonial, anti-racist and black feminists (Alexander, 2005; Hill Collins, 2019; Mohanty & Carty, 2018; Wieringa & Sivori, 2013) have utilised liberatory theories and methodologies to examine co-constituted identities and systems of oppression (not either/or but both) within varying socio-economic, political and cultural terrains both locally and globally. Entangled in the socially constructed identities of gender and sexuality are also spaces of *in between* whereby discursive representations and meanings of the erotic and body are simultaneously embodied and performed into existence. Thus, *queer* or unconventional understandings of being (both pre- and post-modern) incorporate diverse and malleable features that dislodge fixed notions of personhood premised on Enlightenment gender/sex binaries (Allen, 2012; Tinsley, 2008; Wekker, 2006). Through an eclectic theoretical framework, I will problematise the precarity of sexual citizenship for women and LGBTQ persons who are stymied by dominant heteropatriarchal norms. Our intimate and public lives are connected through the state's regulation of sexual conduct, reproduction and family that is embedded in law and the political economy. But marginalised persons are particularly vulnerable in this regard. They have difficulty exercising their erotic autonomy, whether for pleasure, reproduction or profit, because they are exploited and delegitimised by those in power who seek to maintain the status quo.

Feminist and LGBTQ Activism

The theoretical and epistemological considerations of decolonial, feminist and queer scholars in critiquing hegemonies of power are informed by social justice activism in the region that tackles real life issues and seeks to upend injustices through ameliorative measures and by promoting more inclusive and democratic practices in society for the betterment of all. Given the legacy of slavery and colonialism in the Caribbean, there is a rich tradition of social movements in the region defined by a culture of resistance among the downtrodden masses and other oppressed groups who collectively organised to voice their grievances in search of justice. From Pan-Africanism, leftist activism, Marxist-Leninist and trade unionism, black/Caribbean feminisms to transoceanic movements, Boyce-Davies (2013) notes that "a range of intellectual activists, through time, who engage a politics of progressive change from a variety of intersecting political positions" (2013, p. 204). For instance, Caribbean women's realities and activism have always crisscrossed hierarchies of gender, race and class. Under colonialism, they fought against racism and economic deprivation under colonialism alongside their male counterparts in anti-colonial, black nationalist and labour struggles while advocating for universal suffrage, women rights and the betterment of their families and communities through social welfarism and maternal feminist politics. In the post-independence period, women's groups galvanised around improving socio-economic conditions and political rights for women based on basic needs and then layered human rights women/gender and development

(WID/WAD/GAD) approaches during the 1970s to early 1980s.[4] From the mid-1980s into the 1990s, Caribbean feminists vigilantly incorporated a gender analysis in research and public policy initiatives to critically examine unequal power relations between men and women to address gender stereotypes and discrimination that stymie women's advancement in all areas of life (Antrobus, 2004; Barriteau, 2003; Mohammed, 1998; Reddock, 1998). Mainstream Caribbean feminists sought legislative reforms to ameliorate gender discrimination in the law, state and institutions. Their strategies varied in mobilising around violence against women, reproductive rights, fair wages and strengthening women's political leadership. Even with this heightened feminist activism there were some blind spots when it came to the limited representation of Indo-Caribbean women and lesbians in feminist organising and decision-making as well as activists/scholars not fully interrogating the gamut of women's sexual lives in their agenda. In the twenty-first century, a new generation of Caribbean feminist activists are actively exploring the intersections of gender and sexual oppression in advocating for gender and sexual justice for women and LGBTQ persons (Haynes & De Shong, 2017).

[4] Basic needs and women-centered initiatives were ensconced in the women in/and development mandates (WID/WAD) of the United Nations Decade for Women (1975–1985). This was followed by a gender and development agenda influenced by socialist feminist activism led by Development Alternatives with Women for a New Era in 1984. Co-founded by feminist stalwart Peggy Antrobus, this group advocates for the rights of working women and examines the relationship between advanced capitalism (neo-liberalism), gender inequality and sustainable development, particularly of poor and working-class women of colour situated in the global South There was a strong regional response to the WID/WAD/GAD international programme from Caribbean states, agencies and women's groups/feminists. The Women in the Caribbean Project (WICP) 1979–1982 was spearheaded as an empirical research study that examined the everyday lives of Caribbean women in select English-speaking Caribbean countries in the areas of family, work, reproduction and political leadership to name a few (Massiah, 1986). The research findings from WICP that highlighted women's contributions to their communities and the need to break down systemic barriers to fully integrate them in all spheres of society further justified the need for academic programmes in the field, which was initiated by Women and Development Studies Groups on the three campuses of the University of the West Indies in 1982. The institutionalisation of the discipline was cemented with the establishment of the Centre for Gender Studies and Development Studies (now Institute for Gender and Development Studies) in 1993. But this could not have been achieved without the indomitable support and feminist activism from the regional umbrella organisation, Caribbean Association for Feminist Research and Action (CAFRA) which was formed in 1985. CAFRA was founded by Peggy Antrobus, Joan French, Honor Ford-Smith, Sonia Cuales, Rawwida Baksh and Rhoda Reddock.

Therefore, I would like to locate the intensification of LGBTQ activism in the Commonwealth Caribbean in the twenty-first century within the rich tradition of social justice activism in the region. I want to do so because the social experiences of gay, lesbians and transgender persons need to be discussed and analysed in more nuanced ways, beyond surveying their plight in dealing with homophobia and transphobia or providing clinical analysis of their sexual behaviours, to situate them as agents of their lived experiences, producers of knowledge and contributors to civil engagement. Therefore, while it is important to map sexuality studies and examine sexual praxis in the region (Kempadoo, 2009; Sharpe & Pinto, 2006), there is a gap in the scholarship that thoroughly examines the different ways that LGBTQ persons are marginalised as sexual citizens in the Caribbean and how they have mobilised in the fight against discriminatory attitudes, practices and policies in their societies. Their courageous acts of resistance have raised awareness about gender and sexual justice in the public domain in further democratising the societies that they live in. LGBTQ mobilisation in the region has comprised dialogic crisscrossing social justice efforts that include personal acts of resistance by queer youth in the digital age, LGBTQ grassroots organising around HIV/AIDS, human rights, self-help, and capacity building and strategic litigation against discriminatory laws, to name a few (Bulkan & Robinson, 2017; Murray, 2006). Intersectional politics and social networking have also been a part of LGBTQ organising in the region with collaborations with feminists, academics, lawyers, regional and international NGO partnerships and queer diasporic Caribbean communities. In addition to this, the research outputs in the form of technical reports, public policy and scholarly work on Caribbean sexualities, gender and sexuality, same-sex intimacies, sexual health, sex work and the like have contributed to a resource bank that can be tapped into to mitigate injustice.

THE LAW, EPISTEMIC INJUSTICE AND LGBTQ RIGHTS

While scholars have explored sexual citizenship in the Caribbean from legal and socio-cultural perspectives, there has been limited examination of how our intimate, erotic, personal and embodied selves, and relationships with others, get formulated in the production and use of knowledge within the matrices of power in nation-building projects. This book attempts to situate the experiences of LGBTQ persons as sexual citizens in the Commonwealth Caribbean, with a special focus on Barbados, by considering how the prejudice and discrimination that gay, lesbian and transgender persons encounter in state and law produce epistemic injustices and disadvantages. Anti-buggery laws infringe on both the conduct- and identity-based sexual rights of LGBTQ persons. I will examine efforts to abolish anti-buggery laws by uncovering hermeneutical deficiencies in legislation that usurp the sexual rights of LGBTQ persons in language, intent and impact. The subjugated knowledges of LGBTQ persons are credible, and individuals are using their voices to politicise their cause in the region. Moreover, I will draw on the scholarship of Miranda Fricker (2007) and Jose Medina (2013) on epistemic injustice and resistance to investigate how Caribbean LGBTQ persons are disadvantaged when discriminatory laws and state practices infringe on their rights and block resources to them because of their non-dominant gender and sexual identities.

Methodological Considerations and Outline of Chapters

Since I will be investigating how sexual citizenship is circumscribed for Caribbean LGBTQ persons in family, culture, law and state and how they are resisting their oppression, this book comprises an amalgamation of chapters through crisscrossing themes drawing on an eclectic methodology in reporting findings and interpreting data. I use a mixture of primary sources and data (legislation, print media and semi-structured interviews) as well as secondary sources in putting forth a systematic literature review and interpretive analysis in researching gender, sexuality and intimate life. The depth and richness of this research project lends itself to diverse methodologies that are brought together by the epistemic and theoretical commitments regarding the anticipated outcomes of the study.

In Chap. 2, "Sexual Citizenship and the Politics of Exclusion," I carry out a systematic literature review of colonial and post-colonial heteropatriarchal constructions of citizenship that operate through the family, state, law and church to define which sexual citizens are legitimately entitled to rights, resources and recognition in supporting their partnerships and families. This will lay the foundation for me to examine LGBTQ rights in the region in the twenty-first century through their fight for both sexual identity-based rights and sexual conduct-based rights in later chapters. In Chap. 3, "Epistemic Injustice, Social Inequality and LGBTQ Realities," I will examine how LGBTQ persons are strategically negotiating their sexual identity in restrictive, and sometime overtly hostile, social environments because of homophobia and transphobia while simultaneously trying to carve out new temporal and spatial areas to express themselves and lead productive lives. I will analyse the concept of epistemic injustice to shed light on how structural equalities operate to produce testimonial and hermeneutical liabilities via cognitive-affective practices that obscure the social experience of LGBTQ persons from collective understanding. I will also draw on qualitative research from the Human Rights and Sexual Equality Project in Barbados (2014–2015) which centres on the voices of LBGTQ persons derived from semi-structured interviews. In Chap. 4, "The Precarity of Sexual Citizenship: Hermeneutical Injustice, the Law and LGBTQ Rights," the aim is twofold. Firstly, I will examine strategies of resistance employed by LGBTQ groups and activists in advocating for sexual rights by creating counter-knowledges to challenge stigma and discrimination as well as engaging in strategic litigation. Secondly, I will examine the intensification of decriminalisation court cases in the Commonwealth Caribbean, and their success, through critically examining the impact of social justice hermeneutics in upending discriminatory legislation. In Chap. 5, "'It's a Girl Thing': Problematising Female Sexuality, Gender and Lesbophobia in Caribbean Culture," I will further complicate the discourse of sexual rights by interrogating how different subordinated groups have their own power struggles when they do not incorporate an intersectional liberatory praxis to deal with interlocking identities and oppressions of gender and sexuality (sexism and heterosexism), for example, when it comes to the oppression of lesbians or queer women whether in mainstream Caribbean feminism and/or gay activism. This chapter draws on a piece that appeared in the digital collection "Theorizing Homophobia" in 2012. Finally, in Chap. 6, "Decolonising and Queering Caribbean Families," I will explore how the flexibility of African-Caribbean family and kinship networks, and their unconventionally compared to a Eurocentric nuclear familial norm, provides openings for the queering of Caribbean families that allow LGBTQ persons the opportunity to imagine and forge new understandings of intimate-affective relationships that support them and their families. I want to problematise how LGBTQ intimate experiences comprise emotional, affective and epistemic value that is credible. By repositioning Caribbean families and sexualities beyond the heteronormative gaze, meaningful discussion and theorisation of same-sex intimacy will be brought to the fore.

References

Alexander, J. M. (2005). *Pedagogies of crossing: Meditations on feminism, sexual politics, memory, and the sacred.* Duke University Press.

Allen, J. S. (2012). Black/queer/diaspora at the current conjuncture. *GLQ: A Journal of Lesbian and Gay Studies, 18*(2–3), 211–224.

Antrobus, P. (2004). *The global Women's movement: Origins, issues and strategies.* Zed Books.

Barriteau, E. (2003). *Confronting power, theorizing gender: Interdisciplinary perspectives from the Caribbean.* The University of the West Indies Press.

Bulkan, A., & Robinson, T. (2017). Enduring sexed and gendered criminal laws in the anglophone Caribbean. *Caribbean Review of Gender Studies, 11*, 219–240.

Butler, J. (2011). *Bodies that matter: On the discursive limits of sex.* Routledge.

Chateauvert, M. (2008). Framing sexual citizenship: Reconsidering the discourse on African American families. *The Journal of African American History, 93*(2), 198–222.

Cooper, D. (1995). *Power in struggle: Feminism, sexuality, and the state.* New York University Press.

Davies, C. B. (2013). *Caribbean spaces: Escapes from the twilight zone.* University of Illinois Press.

Dixon-Mueller, R., Germain, A., Fredrick, B., & Bourned, K. (2009). Towards a sexual ethics of rights and responsibilities. *Reproductive Health Matters, 17*(33), 111–119(2009).

Epstein, S. (2002). A queer encounter: Sociology and the study of sexuality. In C. L. Williams & A. Stein (Eds.), *Sexuality and gender* (pp. 44–59). Blackwell Publishers.

Foucault, M. (1978). *The history of sexuality. Vol 1: An Introduction.* (R. Hurley, Trans.). Vintage.

Fricker, M. (2007). *Epistemic injustice power & ethics of knowing.* Oxford University Press, Inc..

Glenn, E. N. (2011). Constructing citizenship: Exclusion, subordination, and resistance. *American Sociological Review, 76*(1), 1–24.

Haynes, T., & De Shong, H. A. F. (2017). Queering feminist approaches to gender-based violence in the anglophone Caribbean. *Social and Economic Studies, 66*(1 & 2), 105–131.

Hill Collins, P. (2019). *Intersectionality as critical social theory.* Duke University Press.

Jackson, S., & Scott, S. (1996). *Feminism and sexuality: A reader.* Columbia University Press.

Kempadoo, K. (2009). Caribbean sexuality: Mapping the field. *Caribbean Review of Gender Studies, 3*: 1–24.

Lorde, A. (2017a). Uses of the erotic: The erotic as power. In *A. Lorde your silence will not protect you, anthology* (pp. 22–30). Silver Press.

Lorde, A. (2017b). The transformation of silence into language and action. In *A. Lorde your silence will not protect you, anthology* (pp. 1–6). Silver Press.

Marshall, T. H. (2009). Citizenship and social class. In J. Manza & M. Sauder (Eds.), *Inequality and society* (pp. 148–154). W.W. Norton and Company.

Massiah, J. (1986). Women in the Caribbean Project: An Overview. *Social and Economic Studies, 35*(2), 1–29.

Medina, J. (2013). *The Epistemology of Resistance Gender and Racial Oppression, Epistemic Injustice, and Resistant Imaginations*. Oxford University Press.

Mohammed, P. (1998). Towards indigenous feminist theorizing in the Caribbean. *Feminist Review, 59*, 6–33.

Mohanty, C. T., & Carty, L. (Eds.). (2018). *Feminist freedom warriors: Genealogies, justice, politics and hope*. Haymarket Publications.

Murray, D. (2006). Who's right? Human rights, sexual rights and social change in Barbados. *Culture, Health & Sexuality, 8*(3), 267–281.

Pateman, C. (1988). *The sexual contract* (Kindle ed.). Blackwell Publishing Limited.

Reddock, R. (1998). Women's organizations and movements in the commonwealth Caribbean: The response to the global economic crisis in the 1980s. *Feminist Review, 59*(1), 57–73.

Richardson, D. (2000). *Rethinking sexuality*. Sage.

Robinson, T. (2009). Authorized sex: Same-sex sexuality and law in the Caribbean. In C. Barrow, M. de Bruin, & R. Carr (Eds.), *Sexuality, social exclusion and human rights: Vulnerability in the context of HIV* (pp. 3–22). Ian Randle Publishers.

Sharpe, J., & Pinto, S. (2006). The sweetest taboo: Studies of Caribbean sexualities; A review essay. *Signs: Journal of Women in Culture and Society., 32*(1), 247–277.

Sheller, M. (2012). *Citizenship from below: Erotic agency and Caribbean freedom*. Duke University Press.

Thomas, G. (2007). *The Sexual Demon of Colonial Power. Pan African Embodiment and Erotic Schemes of Empire*. Indiana University Press.

Tinsley, N. O. (2008). Black Atlantic, Queer Atlantic: Queer imaginings of the middle passage. *GLQ: A Journal of Lesbian and Gay Studies, 14*(2–3), 191–215.

Wekker, G. (2006). *The politics of passion: Women's sexual culture in the afro-Surinamese diaspora*. Columbia University Press.

Wieringa, S., & Sivori, H. (2013). *The sexual history of the global south: Sexual politics in Africa, Asia, and Latin America*. Zed Books.

Wilson, A. R. (2009). The 'neat concept' of sexual citizenship: a cautionary tale for human rights discourse. *Contemporary Politics, 15*(1), 73–85.

CHAPTER 2

Sexual Citizenship and the Politics of Exclusion

Abstract Feminists have critiqued androcentric configurations of sexuality that have emerged from patriarchal ideologies and practices that have exploited women's sexuality and denied them sexual independence. Likewise, queer scholars have been critical of the discrimination that LBGTQ persons face from the law based on their sexual expression and orientation due to heterosexism and homophobia. Since Western constructions of sexual citizenship are predicated on racialised, gendered, sexed and classed asymmetries, black/feminist/queer perspectives are critical in deconstructing essentialist notions of gender and sexuality as well as denaturalising sex, in allowing for more diverse and nuanced understandings of sexual activity and intimate life across difference. In this chapter, I want to interrogate how sexual citizenship in the Caribbean is a colonial/post-colonial heteropatriarchal project that has shaped sexual conduct, sexual relations and sexual rights through the family, state and law in legitimising or not legitimising certain racialised, gendered and sexed bodies. In addition to this, I will consider how the existence of diverse sexualities in the Caribbean is critical in denaturalising (queering) and decolonising heteronormative constructions of sexual citizenship and can act as a catalyst for change in LGBTQ resistance and social movements in the region.

Keywords Coloniality, racism and black sexuality • Anti-buggery and
vagrancy law • LGBTQ as "Other" • Nation-building, heteropatriarchy
and sexual citizenship • Redrafting morality in the law • Christian
citizenship • Gender and sexual diversity • Queer Caribbean
subjectivities • Sexual praxis • Sexual rights

> Notions of sexuality are deeply inflected by colonial and imperial inheri-
> tances that have framed nationalism's discourses and silences and continue
> to inform, more or less, the structures of feeling of the region's people
> (Smith, 2011, p. 2)

Sexuality is an integral part of being human based on our erotic experiences with ourselves and others. Sexuality is not neutrally constructed;
instead, it is imbued with generative and consumptive power through intimate relations in the private sphere as well as through its commodification
as socio-sexual capital through the state, institutions and law in the public
sphere. Since what is "sexual and sexualized" is not solely a private affair
(Sheller, 2012, p. 41), sexual citizenship encompasses different aspects of
our intimate lives that have erotic, social and economic value that are
organised and legitimised through the state, law, family and civil society
(Bell & Binnie, 2000; Cooper, 1995; Richardson, 2000; Smith, 2011).
Simply put, "sexuality is not a given, it is a product of negotiation, struggle, and human agency" (Weeks, 2003, p. 19). Sexual citizenship in
Western modernity is characterised by a taxonomy of exclusion that hier-
archises and politicises sexual intimacy based on the interplay of racialised,
gendered and sexed hegemonies within colonial/post-colonial global
capitalism. Part of this history of domination is predicated on coloniality
and intersectional oppressions that reify essentialist and heteropatriarchal
beliefs on gender and sexuality that have been normalised in Caribbean
societies. In *Sex and the Citizen*, Smith (2011) discusses how different
configurations of sexual identities across gender, race, class and nationality
are linked to citizenship through pleasure, production, reproduction and
sex work in categorising who belongs and does not belong to a given
nation, space and time in the Caribbean and its diasporas.

The undoing of a hegemonic notion of sexual citizenship in the region
requires a multifaceted social justice approach that is committed to both
decolonisation and democratisation in bringing about meaningful change
to include all sexual citizens. In this chapter, I will interrogate how sexual

citizenship in the Commonwealth Caribbean is a colonial/post-colonial heteropatriarchal project that has shaped sexual intimacy and rights in the region. I will demonstrate how the precarity of sexual citizenship as derived from a formal citizenship model has shaped intimate life through the regulation of sexual identity and sexual conduct within institutions and culture. I will also consider how the appropriation of erotic resources for pleasure, reproduction or profit, as is the case with women's sexuality, or delegitimising same-sex sexualities as dangerous and unproductive has resulted in the obstruction of LGBTQ sexual rights.

The literature review in this chapter incorporates decolonial, feminist and queer perspectives that interrogate intersecting hegemonies. This will clarify the historical and contemporary configurations of sexual citizenship that have shaped race, gender, sex and class hierarchies in the region. This will also provide the basis to explore the fight for sexual rights for LGBTQ persons in later chapters.

Drawing on this intersectional analysis, I argue that the gendering and heterosexualisation of sexual citizenship in the post-independent period cannot be separated from its racist origins based on the reification of Enlightenment puritan values from a colonial past, which post-emancipated blacks were expected to emulate. In addition to this, I will consider how the existence of diverse sexualities in the Caribbean is critical in denaturalising (queering) and decolonising heteronormative constructions of sexual citizenship and can act as a catalyst for change in LGBTQ resistance and social movements in the region.

COLONISATION, SLAVERY AND PLANTATION SOCIETY

Coloniality, Racism-Sexism and Commodified Black Bodies

The legacy of European imperialism, colonisation, the Transatlantic Slave Trade and chattel slavery has left an indelible mark in the region. European (white), heterosexual, affluent and propertied men were able to elevate themselves as de facto citizens of their nations and overseers of the colonies through enslavement and exploitation of African labour in producing raw materials for industrial capitalist development in the West during between the sixteenth and the mid-nineteenth century (Beckles, 2016; Williams, 1964). Wealthy planters were at the apex of plantation societies that were stratified along race, colour, gender, sexual and class lines. Thus, dominant notions of sexual citizenship during colonialism and thereafter were shaped by a racist heteropatriarchal logic that only included a few. White male colonisers had racialised patriarchal privilege and power over their wives (and all women), their children (free and unfree), and the bodies and sexuality of enslaved black men and women who were their property (Bush, 1989; Shepherd, 1999). In discussing coloniality, empire and erotic schemes, Thomas rightly argues that "it is the super-exploitation of an African labouring force, reproduced in and through sexual violence, which makes modernity possible for Western civilization" (2007, p. 9). Ultimately, white patriarchal capitalists were able to organise their intimate lives in the areas of sex, marriage, family and inheritance since they were recognised and legitimised as citizens, and by extension sexual citizens, under law and state. Enlightenment gender ideologies along with puritan values reinforced asymmetrical gender roles for men and women based on a capitalist nuclear familial model (Barriteau, 2004). Separate sphere ideology relegated white middle-class women to the domestic realm, whereby they were expected to be chaste, pious, dutiful mothers and obedient wives to their husbands. In contrast, their male counterparts not only conferred power as breadwinners and heads of their families, but they were also dominant in political and socio-economic matters in the public domain (Moore & Johnson, 2004).

During slavery, African identity and existence were circumscribed by European domination and bigotry based in anti-blackness. European explorers and scientists between the seventeenth and nineteenth century promoted racist scientific theories and iconography extolling the purity of races with white Europeans on top and black Africans on the bottom (McGruder, 2010). In Barbados, Beckles (2016) notes the "legal definitions of enslaved Africans reflected English perceptions of them as an inferior race whose only role was to provide labour and social services" (p. 26). Since institutionalised violence was normalised, Africans were brutalised and relegated to the realm of subhuman, as chattel or property, while Europeans elevated themselves to the apex of humanity as ultra-rational beings (Thomas, 2007). Missionaries were complicit in giving masters licence over their Africans by professing that "God willed their enslavement" and they were to be tamed to be obedient workers. Otherwise, non-compliance would lead to severe punishment in the afterlife (Beckles, 2016, p. 26). Besides making slaves obedient in work and faith based on racist beliefs, anti-blackness also informed malevolent ideas and attitudes towards black sexuality. Black sexuality and bodies were pathologised as biologically inferior to whites and they were rendered "phenotypically ugly, animalistic, and hypersexual" (Thomas, 2007, p. 4). Thus, white supremacist hatred or fear of blackness meant that masters used different methods of social control to contain and manipulate black sexuality and fecundity.

While colonisers professed discipline and sexual propriety in the public domain, colonial sexual politics was riddled with sexual admixtures and antagonistic sexual relations between the oppressor and the oppressed along race, colour, gender and class lines (Kempadoo, 2003; Mohammed, 2000). As property, slaves did not have full bodily and erotic autonomy to make decisions about intimate relations and reproduction. Race and gender oppression contributed to slave women being exploited for their productive and reproductive labour in the output of raw materials and offspring for the plantocracy as well as for fulfilling the sexual desires of white men. Beckles (2003) notes that "in the laws of the island during the 17th and 18th centuries, a man could not rape his slave; the slave had neither legal rights nor personal identity, and masters could do as they wished with slaves" (pp. 142–143).

While white men terrorised black women with impunity and shamelessly hid their paternity of illegitimate slave children, black women had to grapple with their victimisation alongside motherhood. The sexual commodification of non-white female bodies did not end there. Some white men forced their coloured mistresses and black women into prostitution for economic gain in urban centres (Beckles, 2003). Kempadoo (2003) argues that the foundation of prostitution and sex tourism in the Caribbean was a colonial operative that reified the colonisers' desire to control and claim the sexuality of all women (white, coloured and black). Based on racist and sexist stereotypes, white women epitomised true femininity and beauty while brown and black counterparts were lowly ranked respectively on the social hierarchy.

Post-Emancipation Period: Social Control and Vagrancy and Anti-Buggery Laws

The abolishment of slavery in the British colonies in 1834 did not bring the end of suffering for black people. In the post-emancipation period, ex-slaves struggled to make ends meet under the weight of colonial dominance, racism and poverty. A defective plantation economy had little to offer them besides low-paying menial jobs in agriculture and service work. Colonised African-Caribbean people had limited rights and opportunities for upward mobility, and racist fears about their bodies had not abated.

Instead, their racialised, sexed and gendered bodies were policed to keep them compliant and further inculcate them into the colonial Eurocentric status quo. Sheller (2012) argues that "thus, racial, ethnic, gendered, and sexual claims to citizenship in the post-slavery Caribbean emerge as attempts to institute specifically embodied masculinities and femininities that are always in tension with state efforts to discipline sexuality, fertility and labour relations" (p. 27). The precarity of sexual autonomy and rights ensued for black masses, which included the debasement of the flexible and adaptive features of African-Caribbean families and heterosexual socio-sexual unions, such as polyamory, visiting relationships, matrifocality and extended family units. These features were characterised as deviant and dysfunctional compared to an Enlightenment gender ideology that privileged marriage, nuclear family and male headship.

To perpetuate this system, Robinson (2017) argues that nineteenth- and early twentieth-century laws were put in place to delegitimise the status of children born outside of marriage to protect conjugal paternity. This was done by creating a new category of woman ('single woman'—*read single working-class black woman*) in opposition to the respectable married woman who had to petition the court to gain economic assistance from the state and negligent fathers. Not surprisingly, the delegitimisation of black sexuality and families coincided with intense assimilationist socio-cultural and religious strategies of the empire to refine the behaviour and culture of the black masses through "Christian morality and Victorian etiquette" to pacify them and to keep them from rebelling against white authority (Moore & Johnson, 2004, p. 137). In post-slavery Jamaica, upwardly mobile black and coloured elites sought "to adopt a new but uncomfortable *gentilité* which may have marked them as approaching the civilized ways of their middle-class cousins in England, but which they wore as an ill-fitting cloak" (ibid., p. 143). Thus, Christian colonial education was used as a corrective measure to usher blacks into civility and save them from ungodliness through academics, religious instruction and heteronormative gender roles. Colonial education stressed separate sphere ideology and the education girls received was linked to arts, culture and household management preparing them to become *good* wives and mothers, while boys were streamed into math, science and technical courses to mould them into thinkers, disciplinarians and providers (Reddock, 1994). The gender and sexual politics at the time demanded that women lead chaste and wholesome lives as caregivers and carriers of culture so sex for procreation, and not pleasure, was the order of the day.

Other disciplinary methods used to police and pacify poor and working-class black people and further delegitimise their claims to citizenship were instituted through vagrancy and anti-buggery laws. The aim was to diminish Afro-creole socio-cultural expression and queer sexuality that were seen as threats to the colonial heteropatriarchal order. Alexander (2013) notes that vagrancy and anti-obeah laws restricted the mobility and cultural expressions of poor and working-class black people in urban centres through prohibitions against drumming, stick fighting, carnival and the practice of African-derived religious traditions. Additionally, the racist suppression of black creole culture coalesced with heteronormative norms through the increased policing of homosexuality via anti-buggery laws during the 1860s. Thus, the intersectionality of blackness and homosexuality was doubly deviant. Gaskins (2013) adds that "it was in the final decades of the 19th century that outright hostility toward homosexual acts became common, specifically during the Victorian era. Anxiety about homosexuality was fuelled by fears of declining middle-class values and perceived threats to the British Empire" (p. 431).

The first Buggery Act came into existence in 1533 under the reign of King Henry VIII in England in deeming anal sex, bestiality and other non-procreative sexual acts as unnatural and therefore sinful sexual acts against God and man. The initial punishment was death; this was later changed to life imprisonment. With the rise of British colonialism and participation in the Transatlantic Slave Trade during the eighteenth and nineteenth centuries, Jackman (2016) confirms that "Britain exported its views on sexuality to its colonies and so, these countries were subjected to the 1861 Offences to the Person Act–which carried a penalty of imprisonment for the 'abominable crime of buggery' (that, is, anal sex)–and later, the 1885 Criminal Law Amendment Act, which introduced penalties for acts of 'gross indecency' between men" (p. 131). British authorities took a big stick approach to the colonies due to their distance from the metropole and to quell delinquent and rebellious behaviour. Gaskins (2013) contends that white men in the colonies may have had more freedom and opportunity to engage in same-sex relations if they desired to because they were away from the prying eyes of their families and society and also due to the fact that during initial phases of colonial migration "British colonisers lived in an almost all-male society with few outlets for heterosexual sex and with little legal restrictions" (p. 431). In fact, European travellers' logs disparagingly referenced the islands as a place of sodomites, which is a biblical term for gay men (ibid.). Although there is a paucity of research on same-sex intimacies in slave communities, there is evidence of same-sex eroticism and cross-gender representations in Afro-creole spirituality and cultural art forms such as carnival in the post-emancipation period (King, 2014; Wekker, 2009). Bulkan and Robinson surmise that "a body of raced, gendered and sexed post-slavery criminal laws, and their legal constructions of deviance and conceptions of punishment, shadow the modern Caribbean" (2017, p. 221). Since black creole freedom brought with it personal and collective acts of resistance, which included the visibility of black queer bodies in the public domain, all attempts were made to suppress recalcitrant behaviour that defied colonial supremacy.

NATION-BUILDING, HETEROPATRIARCHY AND SEXUAL CITIZENSHIP

English-speaking Caribbean nations were ushered into independence during the 1960s and the period brought with it sovereignty as well as economic, educational and political nation-building initiatives to improve the welfare of the general population. While Caribbean people were liberated from the physical shackles of the colonial masters, ideological impediments persisted due to the idealisation and internalisation of bourgeois norms and values by the black elite who came to power (Kamugisha, 2019). The gendering of citizenship was operationalised through the law, state and family within dependent capitalist economies during the post-independence period. The 1975–1985 UN Decade for Women ushered in policies and practices based on a Women and Development (WAD) model, and Gender and Development (GAD) thereafter, that sought state and NGO interventions to mitigate economic, educational, political and domestic hardships that compromise their rights, well-being and full integration into mainstream society (Antrobus, 2004). This liberal development model was critiqued by feminists who sought a more progressive politicisation of women's grievances beyond a framework of basic needs (food, shelter, healthcare and education) and traditional division of labour. Instead, they fought for legislative and institutional changes to combat gender stereotypes, discrimination, poverty, pay inequality and violence that disadvantaged women in their lives and stalled their upward mobility.

Nation-building and modernising initiatives were guided by Enlightenment middle-class gender ideology that constructed modern

Caribbean men and women through asymmetrical gender roles in the family and society. Barriteau (2004) notes that heterosexual women were integrated into society based on utilitarian principles whereby women were seen as stewards of the nation in helping with national development and in sacrificing for the common good. She goes on to argue that secondary citizenship status assigned to women is based on asymmetrical gender relations that operate both ideologically and materially in defining women as semi-autonomous beings compared to men. Men benefit from male privilege and, thus, they are granted "a priori rights" as citizens (p. 439). From a feminist legal perspective, Robinson (2003) examines in "Beyond the Bill of Rights: Sexing the Citizen" the contradictions of citizenship for women. She adds that while there is a symbolic recognition of gender equality in the constitutions of some Caribbean nation-states, the application of rights is not equally weighted since "men remain the paradigm of a citizen and, insignificant measure, women are included as citizens through their relationship to men" (p. 232). Discriminatory laws and practices placed conditionalities on citizenship for women which had ramifications for their sexual and intimate lives. Robinson states that a woman could not pass on citizenship to her foreign-born husband since "men are seen as giver of rights," and there were legal exemptions for marital rape and unmarried pregnant teachers being fired because they did not represent respectable womanhood (ibid., p. 236). So, the mandates of gender inclusion in nation-building were reformist because they relied on respectability politics and gender stereotypes that subordinated women to men through the "good woman of the nation" trope. It also had another blind spot: women were decisively constructed as heterosexual within the nation. This was also an oversight by some Caribbean feminists who engaged in gender essentialism and presented women as an undifferentiated oppressed group in society while critiquing male dominance. In considering the intersectionality of gender, sexuality and post-coloniality, Alexander argues that "heteropatriarchal recolonization operates through the consolidation of certain psychic economies and racialised hierarchies as well as within various material and ideological processes initiated by the state, both inside and beyond the law" (Alexander, 2005, p. 26). The next section will illustrate how sexual citizenship for Caribbean women has been shaped by both gender and heteronormative idealisations that deny their sexual agency and punish them, especially lesbians, single mothers and sex workers, for contravening the status quo.

The Law, Heteronormativity and LGBTQ Marginalisation

While it was purposeful to partially incorporate heterosexual women into the nation as good wives and mothers and carriers of culture, LGBTQ people were outrightly threatening to the status quo. Alexander (1994), Robinson (2009) and Tambia (2011) from varying perspectives discuss how Caribbean governments strategically redrafted morality into their criminal codes to restrict same-sex activity through stiffer penalties for buggery. Robinson notes that amendments to the Sexual Offences laws during the late 1980s "significantly refined the definition of buggery or the unnatural crime and increased the severity of the punishment" (Robinson, 2009, p. 12). So, by the 1990s, "whom you had sex with" was just as important as the "sex you did" in classifying legitimate and non-legitimate sexual acts and citizens (Robinson, 2009, p. 12). The heteropatriarchal recolonisation of the law was entangled with religious doctrine advocated for by the church and religious civil society organisations that propagated Christian citizenship (Lazarus, 2015). Usually, gay men are the target of punitive laws and public disdain but the heightened wave of conservative moralism at this time meant that lesbian and queer women were also under attack. Alexander (1994) notes in "Not Just (Any) Body Can Be a Citizen" that consensual sexual activity between women became a criminal offence in Trinidad and Tobago and the Bahamas in the late twentieth century. She goes on to argue that lesbians or women who have sex with women (WSW) were targeted because they were perceived as contravening procreative heterosexual activity within the nuclear family, and cannot be readily appropriated by men and the state for economic gain. In a later work, "Erotic Autonomy as a Politics of Decolonization: Feminism, Tourism, and the State on the Bahamas," she considers sexual agency for women as a self-directed sexual freedom or governance or "erotic autonomy" that can be harnessed to resist oppression. Alexander professes that "erotic autonomy signals danger to the heterosexual family and to the nation. And because loyalty to the nation as a citizen is perennially colonized within reproduction and heterosexuality, erotic autonomy brings with it undoing the nation entirely, a possible charge of irresponsible citizenship, or no responsibility at all" (Alexander, 2005, p. 23). I will revisit this notion of erotic autonomy in later chapters when I examine LGBTQ resistance against discriminatory laws and practices that infringe on their sexual rights. While some have questioned the efficacy of anti-buggery laws because they are rarely enforced between consenting adults, nevertheless, they are punitive because "they all give social and legal sanction for discrimination, violence, and prejudice against lesbian, gay, bisexual and transgender (LGBTQ) individuals" (Human Rights Watch, 2017, p. 9). Acts of violence against LGBTQ persons vary and include them being harassed, threatened, physically and sexually assaulted, and even murdered. Public violence can be exacerbated by state and institutional violence whereby the police do not take the victimisation of LGBTQ persons seriously when reported (Gaskins, 2013). Transgender persons face additional vulnerability due to their non-conforming gender identity and expression (Crawford, 2019). Thus, violence tends to be either minimised or under-reported in the print media (Haynes & De Shong, 2017). Moreover, the precarity of sexual citizenship for LGBTQ persons means that their embodied affective and libidinal expressions and interactions are unfairly devalued and treated compared to heterosexual persons, leaving them at the margin of their societies.

Finally, there have been different occasions where politicians have been pressured by LGBTQ activists to address the lack of equal protection and rights before the law for gay, lesbian and transgender individuals. The responses have been either half-hearted or have reflected their unwillingness to challenge discriminatory laws and practices that impact the lives of sexual minorities.

DECOLONISING AND QUEERING SEXUAL CITIZENSHIP

Given the colonial legacy in the region, sexual citizenship in the Caribbean must be interrogated through a decolonial, queer and feminist intersectional lens to challenge a post-colonial logic that privileges heterosexuality and continues to characterise queer sexualities as deviant and morally corrupting to society. It is necessary to deconstruct Western positivist ways of being that have naturalised and reduced gender and sexuality to the biological or a binary to be universally applied to all human beings without considering diverse and polyvariant representations of gender and sexuality that have existed, and exist, cross-culturally throughout time and space (Weeks, 2003; Wieringa & Sivori, 2013; Guadio, 2001). Caribbean scholars and writers have provided a rich tapestry of work that humanises queer creole identities and socio-sexual relationships as they navigate both the joys and trials of life in Caribbean and Caribbean diasporic locations (Glave, 2008). In black Queer Studies, different aspects of blackness and queerness have been explored through diaspora and hybridity based on voluntary and involuntary movements in the black Atlantic experience (Allen, 2012; Tinsley, 2008). Elsewhere, the term queer *has been troubled* as not being something new in the region by documenting the unconventionality, fluidity and diversity of Caribbean sexualities as creolised formations (King, 2014) as "a kind of scholarly disidentification" (Haynes & De Shong, 2017, p. 108) to destabilise Western epistemic ownership of it. In addition to this, given the different cultural creolised configurations that have shaped gender and sexual identities and expressions in the region, alongside transnational social formations, and hybrid identities in the Caribbean diaspora across borders and generations, there is no escaping the multiple (and sometimes contradictory) ways Caribbean people choose to identify and name themselves. As a result, *queer* will remain a contested term in the region. Therefore, I am more interested in *queering* as a process of deconstructing hegemonic categories of gender and sexuality rather than using it to establish truth claims about sexual identity and naming in the region.

The Caribbean is not a monolithic region especially when it comes to culture, gender and sexuality and it should not be presented as such. It is a racially stratified poly-cultural/ethnic/linguistic space and its creolised geographies house diverse gender and sexual expressions and arrangements, both named and unnamed, that intersect with hegemonic norms (Sheller, 2012). In mapping sexualities in the Caribbean, Kempadoo (2009) surveys the interplay of conventional and unconventional sexual relations in a dynamic Caribbean sexscape underpinned by coloniality and creolised configurations. She states that sexuality in the Caribbean "is characterised by patriarchal heteronormativity yet includes bisexual and same-sex relations. It is powerful or violent, frequently acts as an economic resource, sustains polygamy, multiple partnering, and polyamory, and is mediated by constructions of race, ethnicity, and racism" (ibid., p. 12). In further elucidating the existence of diverse sexual formations in the Caribbean that are fluid and may predate modern categorisations, Wekker (2009) captures a creole entanglement of gender, sexuality and spirituality as it relates to erotic autonomy of working-class Afro-Surinamese women. These women practise the polytheistic African-derived religion of Winti that allows them to be sexually open and creative in being in tune with nature, others and themselves. Through *mati work* women forge supportive and pleasurable partnerships with other women that are social, sexual or both, even alongside their relationships with men. In looking at gender, sexuality and performativity in culture, Rosamond

sexualities" (Bell & Binnie, 2000, p. 10). Sexual identity, like gender identity, is embodied as well as subjectively shaped by personal experiences and societal norms. No one is exclusively outside sexual identification within the matrices of power. Thus, we should not minimise the "connection between sexuality, embodiment and politics" (Sheller, 2012, p. 39). The way a person chooses to identify or not depends on which identity they decide to privilege over others based on how they see themselves, their relationships with others and societal expectations. For instance, some bisexual men who adhere to hegemonic masculinity (Lewis, 2003) may value and identify with the social capital and privilege associated with heterosexuality regardless of who they are sleeping with behind closed doors. In this case, they are presumed to be heterosexual and may identify as such based on dominant social norms; hence, sexual identification is still taking place. Kempadoo's second reason for focusing on a sexual praxis is that "the specification of sexual identity groups often elides the very varied sexual arrangements in the region and can work to hinder broader understandings of how Caribbean peoples relate sexually" (Kempadoo, 2009, p. 2). This seems to suggest that "sexual identity groups" (assuming that this is a reference to LGBTQ groups) may occlude the diverse sexual arrangements in Caribbean societies. Conversely, I would argue that the politicisation of LGBTQ rights strengthens our understanding of diverse sexual arrangements in the Caribbean, both named and unnamed, within unequal power relations that are not captured by behavioural explanations. This reiterates the importance of conduct-based and identity-based sexual rights being important in challenging stigma and discrimination in the region. Thus, LGBTQ activism in the fight for sexual rights has been sparked by personal defiance, civil society engagement, strategic litigation and Caribbean diasporic alliances to expand parameters of citizenship for a new generation of sexual citizens.

In conclusion, sexual citizenship in the Caribbean has been shaped by colonial and post-colonial discourses and practices rooted in heteronationalism which cannot be separated from the gendered and racialised exploitation experienced by black subaltern bodies. The marginalisation of women and LGBTQ persons as sexual citizens is not accidental; rather, it is purposeful in the state using its power to regulate libidinal desires and reproduction to suit its productive and consumptive needs. The debasement of homosexuality through moralistic attitudes by black and brown elites and religious zealots reflects internalised oppression and how the emancipation of black people was carefully monitored in public as well as in private to ensure that intimate life mirrored their colonial predecessors as a marker of modernity and to protect the socio-economic and political interests of those in power. The racial liberation of black Caribbean people was incomplete because it did not seek to overturn Eurocentric cultural values that were entangled in colonial life in the areas of gender, sexuality and family, and how black creoles were inculcated into this worldview which called for them to elevate themselves as sexual citizens based on bourgeois norms. The suppression of African creole culture and erasure of polyvariant sexualities are indicative of the exclusionary politics of sexual citizenship that have been normalised in the state and law. So, it is not surprising that LGBTQ rights are being championed at this current time given the rich history of resistance and social movements in the region.

REFERENCES

Alexander, J. M. (1994). Not just (any)body can be a citizen: The politics of law, sexuality and post-coloniality in Trinidad and Tobago and Bahamas. *Feminist Review, 48,* 5–23.

Alexander, J. M. (2005). *Pedagogies of crossing: meditations on feminism, sexual politics, memory, and the sacred.* Duke University Press.

Alexander, J. M. (2013). *Decolonisation as healing practice: The unfinished project of (Caribbean) feminism. The 19th annual Caribbean women catalysts for change public lecture.* The University of the West Indies, Cave Hill Campus.

Allen, J.S (2012). Black/Queer/Diaspora at the Current Conjuncture. *GLQ: A Journal of Lesbian and Gay Studies, 18*(2–3), 211–224.

Antrobus, P. (2004). *The global women's movement: origins, Issues and Strategies.* Zed Books.

Barriteau, E. (2004). Constructing feminist knowledge in the commonwealth Caribbean in the era of globalization. In B. Bailey & E. Leo-Rhynie (Eds.), *Gender in the 21ˢᵗ century: Caribbean perspectives, visions and possibilities* (pp. 437–465). Ian Randle Publishers.

Beckles, H. M. (2003). Perfect property: Enslaved black women. In E. Barriteau (Ed.), *Confronting power theorizing gender: Interdisciplinary perspectives in the Caribbean* (pp. 142–158). The University of the West Indies Press.

Beckles, H. M. (2016). *The first black slave society Britain's Barbarity in Barbados* (pp. 1636–1876). The University of the West Indies Press.

Bell, D., & Binnie, J. (2000). *The sexual citizen: Queer politics and beyond.* Polity Press.

Bulkan, A., & Robinson, T. (2017). Enduring sexed and gendered criminal laws in the Anglophone Caribbean. *Caribbean Review of Gender Studies Issue, 11,* 219–240.

Bush, B. (1989). *Slave Women in Caribbean Society 1650–1838*. Bloomington: Indiana University Press.

Butler, J. (2011). *Bodies that matter: On the discursive limits of sex*. Routledge.

Cooper, D. (1995). *Power in struggle: feminism, sexuality, and the state*. New York University Press.

Crawford, C. (2019). Unbearable knowledge: sexual citizenship, homophobia and the taxonomy of ignorance. *Journal of Eastern Caribbean Studies, 44*(2), 115–144.

Dixon-Mueller, R., Germain, A., Fredrick, B., & Bourned, K. (2009). Towards a sexual ethics of rights and responsibilities. *Reproductive Health Matters, 17*(33), 111–119.

Epstein, S. (2002). A Queer encounter: Sociology and the study of sexuality. In C. L. Williams & A. Stein (Eds.), *Sexuality and Gender* (pp. 44–59). Blackwell Publishers.

Gaskins, J. (2013). 'Buggery' and the commonwealth Caribbean: a comparative examination of the Bahamas, Jamaica, and Trinidad and Tobago. In C. Lennox & M. Waites (Eds.), *Human rights, sexual orientation and sex identity in the commonwealth: Struggles for decriminalization and change* (pp. 429–454). University of London Press.

Glave, T. (2008). *Our Caribbean: A Gathering of Lesbian and Gay Writing from the Antilles*. Duke University Press.

Guadio, R. P. (2001). Male lesbians and other queer notions in Hausa. In S. Murray & W. Roscoe (Eds.), *Boy-wives and female husbands: Studies of African homosexualities* (pp. 115–126). Palgrave Macmillan.

Haynes, T., & De Shong, H. A. F. (2017). Queering feminist approaches to gender-based violence in the anglophone Caribbean. *Social and Economic Studies, 66*(1 & 2), 105–131.

Human Rights Watch. (2017). *I have to leave me behind: Discriminatory laws against LGBT people in the Eastern Caribbean* report. United States. https://www.hrw.org/report/2018/03/21/i-have-leave-be-me/discriminatory-laws-against-lgbt-people-eastern-caribbean

Jackman, M. (2016). They called it the 'abominable crime': An analysis of heterosexual support for anti-gay laws in Barbados, Guyana and Trinidad and Tobago. *Sex Res Soc Policy, 13*, 130–141. https://doi.org/10.1007/s13178-015-0209-6

Kamugisha, A. (2019). *Beyond coloniality: Citizenship and freedom in the Caribbean intellectual tradition*. Indiana University Press.

Kempadoo, K. (2003). Theorizing sexual relations in the Caribbean: Prostitution and the problem of the "Exotic". In E. Barriteau (Ed.), *Confronting power, theorizing gender: Interdisciplinary perspectives from the Caribbean* (pp. 159–185). The University of the West Indies Press.

Kempadoo, K. (2009). Caribbean sexuality: Mapping the field. *Caribbean Review of Gender Studies, 3*, 1–24. http://sta.uwi.edu/crgs/november2009/journals/Kempadoo.pdf

King, R. (2011). New citizens, new sexualities: Nineteenth century jamettes. In F. Smith (Ed.), *Sex and the citizen—interrogating the Caribbean* (pp. 214–224). University of Virginia Press.

King, R. (2014). *Island bodies: Transgressive sexualities in the Caribbean imagination.* University of Florida Press.

Lazarus, L. (2015). Sexual citizenship and conservative Christian mobilisation in Jamaica. *Journal of Eastern Caribbean Studies, 40*(1), 109–140.

Lewis, L. (2003). *The Culture of gender and sexuality in the Caribbean.* University Press of Florida.

McGruder, K. (2010). Pathologizing black sexuality. In J. Battle & S. L. Barnes (Eds.), *Black sexualities: Probing powers, passions, practices and politics* (pp. 101–118). Rutgers University Press.

Mohammed, P. (2000). But most of all mi love me browning: The emergence in eighteenth and nineteenth century Jamaica of the mulatto woman as the desired. *Feminist Review, 65*(1), 22–48.

Mohammed, R. (2017). B-GLAD voices: The LGBT country report. Barbados, Gays, Lesbians and All-Sexuals against Discrimination. https://www.academia.edu/36091654/VOICES_-_Barbados_LGBT_Country_Report

Moore, B. L., & Johnson, M. A. (2004). Manners maketh (Wo)man': Transforming the Jamaican character. In B. L. Moore & M. A. Johnson (Eds.), *Neither led nor driven: Contesting British Cultural Imperialism in Jamaica, 1865–1920* (pp. 137–156). The University of the West Indies Press.

Murray, D. (2006). Who's right? Human rights, sexual rights and social change in Barbados. *Culture, Health & Sexuality, 8*(3), 267–281.

Murray, D. (2009). Bajan queens, nebulous scenes: Sexual diversity in Barbados. *Caribbean Review of Gender Studies, 3*, 1–20.

Reddock, R. (1994). *Women, Labour and Politics in Trinidad and Tobago.* Zed Books.

Richardson, D. (2000). *Rethinking sexuality.* Sage.

Robinson, T. (2000). Fictions of citizenship, bodies without sex: The production and effacement of gender in law. *Small Axe, 7*, 1–27.

Robinson, T. (2003). Beyond the bill of rights: sexing the citizen. In E. Barriteau (Ed.), *Confronting power, theorizing gender: Interdisciplinary perspectives from the Caribbean* (pp. 231–261). UWI Press.

Robinson, T. (2009). Authorized sex: Same-sex sexuality and law in the Caribbean. In C. Barrow, M. de Bruin, & R. Carr (Eds.), *Sexuality, social exclusion and human rights: Vulnerability in the context of HIV* (pp. 3–22). Ian Randle Publishers.

Robinson, T. (2017). Valuing care work. *Journal of Eastern Caribbean Studies.,* 42(3), 60–79.

Sheller, M. (2012). *Citizenship from below: Erotic agency and Caribbean freedom.* Duke University Press.

Shepherd, V. (1999). *Women in Caribbean history.* Ian Randle Publishers.

Smith, F. (2011). *Sex and the citizen: Interrogating the Caribbean.* University of Virginia Press.

Tambia, Y. (2011). Threatening sexual (Mis) behavior homosexuality in the penal code debates in Trinidad and Tobago 1986. In F. Smith (Ed.), *Sex and the citizen: Interrogating the Caribbean* (pp. 143–156). University of Virginia Press.

Thomas, G. (2007). *The sexual demon of colonial power.* Indiana University Press.

Tinsley, N. O. (2008). Black Atlantic, Queer Atlantic: Queer imaginings of the middle passage. *GLQ: A Journal of Lesbian and Gay Studies, 14*(2–3), 191–215.

Wekker, G. (2009). Afro-surinamese women's sexual culture and the long shadows of the past. In C. Barrow, M. de Bruin, & R. Carr (Eds.), *Sexuality, social exclusion and human rights: Vulnerability in the context of HIV* (pp. 192–213). Ian Randle Publishers.

Weeks, J. (2003). *Sexuality.* Second Edition. London: Routledge.

Wieringa, S., & Sivori, H. (2013). *The sexual history of the global south: Sexual Politics in Africa, Asia, and Latin America.* Zed Books.

Williams, E. (1964). *Capitalism and slavery.* Andre Deutsch.

CHAPTER 3

Epistemic Injustice, Sexual Inequality and LGBTQ Realities

Abstract In this chapter, I will examine the central points on epistemic injustice by Miranda Fricker ((2007). *Epistemic Injustice Power & Ethics of knowing*. Oxford University Press, Inc.) and Jose Medina ((2013). *The epistemology of resistance: Gender and racial oppression, epistemic injustice, and the social imagination*. Oxford University Press.) and relate them to the experiences of LGBTQ persons in the Commonwealth Caribbean. I will examine how testimonial injustice impacts their ability to be heard as credible subjects/speakers in their own right because of the silences and disadvantages that they experience in their communities and wider society. I will also consider how LGBTQ persons must navigate multiple, and oftentimes contradictory, cognitive-affective feelings, emotions and social cues to mitigate isolation, harassment and violence in their lives because of their non-conforming gender and sexual identities. The personal angst and trauma caused by homophobia and transphobia requires personal awareness and resiliency to challenge and overcome oppression. Therefore, I will examine the different ways LGBTQ persons are using their voices and resistant knowledges to break silences as well as to decrease communicative and interpretive deficits between them and those who seek to maintain the status quo to their detriment.

Keywords Oppression and epistemic Injustice • Language and testimonial injustice • Semiotics and hermeneutical injustice • Heteronormativity, homophobia and transphobia • Caribbean LGBTQ counter-knowledges

Pedagogies summons subordinated knowledges that are produced in the context of the practices of marginalization in order that we might destabilize existing practices of knowing and thus cross the fictive boundaries of exclusion and marginalization. This, then, is the existential message of the Crossing—to apprehend how it might instruct us in the urgent task of configuring new ways of being and knowing and to plot the different metaphysics that are needed to move away from living alterity premised in difference to living intersubjectivity premised in relationality and solidarity. (Alexander, 2005, pp. 7–8)

In *Pedagogies of Crossing: Meditations on Feminism, Sexual Politics, Memory, and the Sacred*, Alexander (2005) considers the different ways that women and sexual minorities are engaged in the creation of new knowledge and ways of being through liberatory practices that help them validate themselves and connect them to others. While marginalised groups must continuously struggle against oppressive forces, Alexander does not equate the struggle in and of itself as the defining characteristic of the knower, nor does she reduce the identity, knowledge and actions of the knower to rigid identity politics whereby an "alterity premised in difference" is closed off from others or is seen as a fait accompli" (2005, p. 8). Alexander's insights on subjugated gender and sexual identities and knowledges, and their relationship to power, resistance and transversal politics, provide an opening to discuss the relationship between sexual citizenship and epistemic injustice. Although scholars have explored sexual citizenship in the Caribbean from legal and socio-cultural perspectives, there has been limited examination of how our intimate, erotic, personal and embodied selves, and relationships with others, get formulated in the production and use of knowledge, and by extension language, within the matrices of power in social relations and the nation-building project.

In considering Alexander's examination of racialised, gendered and queer "subjugated knowledges" (2005, p. 7) and their crossings within global capitalism, in this chapter I want to draw on Miranda Fricker's (2007) work on epistemic injustices and structural inequalities to investigate the experiences of Caribbean LGBTQ persons with homophobia and transphobia in the Caribbean. In *Epistemic Injustice: Power and the Ethics of Knowing* (2007), Fricker uses a critical philosophical lens to examine the different ways unfairness can take place when the knowledge of the knower is not valued because of who they are (woman, black, gay, disabled and the like). For Fricker, epistemic justice must be a part of democratic principles and practices to challenge social inequality and to bring people together. She states that "the overarching aim is to bring to light certain ethical aspects of two of our most basic everyday epistemic practices: conveying knowledge to others by telling them and making sense of our own social experiences" (2007, p. 1). Fricker goes on to address how social inequalities are produced and sustained through two types of epistemic injustice (testimonial and hermeneutical) and why it is necessary to overcome them to challenge prejudice and discrimination against marginalised groups as well to democratise institutions and social relations. In addition to this, José Medina, in *The Epistemology of Resistance: Gender and Racial Oppression, Epistemic Injustice, and the Social Imagination* (2013), provides a thorough examination of Fricker's framework of epistemic injustice. The author simultaneously critiques and builds on Fricker's scholarship in concretising the different types of epistemic injustice in relation to race and gender oppression and discusses the agency that marginalised groups can harness by engaging in epistemic resistance.

PROBLEMATISING EPISTEMIC INJUSTICE

Fricker (2007) is concerned that epistemic injustice can harm the self-worth of individuals. Oppressive situations may produce adverse emotional and psychological effects that prevent individuals from recognising that they are intrinsically valuable to themselves and others. Epistemic justice is vital to human life in how knowledge is acquired and shared among people in bringing them together. However, when empathy is lacking in a relationship an individual can be wronged "in their capacity as a subject of knowledge" (2007, p. 5). She refers to the notion of "identity power" (2007, p. 14) to denote the different ways that identities are marked and understood by individuals, as speaker and hearer, through unequal power relations. Due to the historical and contemporary structural inequalities, socially constructed identity standpoints are not neutrally configured in relation to *who speaks for whom* or *who listens to whom* because they exist within a matrix of domination (Hill Collins, 2005). For instance, decolonial, transnational and black feminist theorising and activism have brought to light how intersectional or co-constituted identities and oppressions inform the realities of black women and women of colour across differences, and how feminists and queer activists of colour have challenged local and global racialised and gendered heteronormative hegemonies (Hill Collins, 2005; Mohanty & Carty, 2018). For Fricker, an abuse of power and epistemic harms go hand in hand. She challenges a postmodernist viewpoint that sees power as being "socially disseminated" or distanced from (2007, p. 10) situated knowers. While Foucault focuses on the structures or institutions of power which are acted upon the individual, Fricker argues that power is both hierarchal and relational and it is "dependent upon co-ordination with social others, are in a sense socially situated" (2007, p. 11). Hence, she does not view oppressive systems operating in isolation from the unequal social relations that create and reproduce them.

Fricker (2007) goes on to discuss two types of epistemic injustices: testimonial and hermeneutical. These injustices occur when an individual is not listened to or not understood because of their identity. The first type of epistemic injustice is testimonial injustice, whereby the speaker's credibility is questioned or the speaker is not *heard* because of who they are. Fricker states that "the basic idea is that a speaker suffers a testimonial injustice just if prejudice on the hearer's part causes him to give the speaker less credibility than he would otherwise have given" (2007, p. 4). Sher gives the example of an individual being perceived as criminal by police because they are black. While Fricker notes that the premise behind testimonial injustice is that identity prejudice takes place when the hearer prejudges the speaker because of some aspect of their identity, it cannot be separated from the ideological and material basis that created the oppression in the first place, which leads us to the second type of epistemic injustice: hermeneutical injustice.

Fricker states that hermeneutical injustice occurs when someone's "social situation is such that a collective hermeneutical gap prevents them in particular from making sense of an experience which it is strongly in their interests to render intelligible" (2007, p. 7). On a personal level, hermeneutical injustice can arise when an individual lacks the interpretive resources and knowledge to overcome their oppression, resulting in feelings of helplessness and powerlessness. Hermeneutical injustice can also occur from an oppressor's position whereby an individual commits an egregious act against someone but fails to acknowledge the infraction and change their behaviour. Their prejudicial attitude may be a result of ignorance (lack of knowledge) and/or arrogance based on their privileged status in society. Hermeneutical injustice also operates on a structural level when marginalised groups experience systemic discrimination because their "social experience [is] obscured from collective understanding owing to a structural identity prejudice in the collective hermeneutical resource"

(2007, p. 155). Fricker notes that hermeneutical injustice on a structural level is difficult to overcome because knowledge and resources are connected to hegemonic power. So, they are not equitably distributed among individuals. In linking social justice with ethical practices, Fricker contends that hermeneutical virtue is an important principle in advocating for social change because ethical behaviour and having empathy for others may help an individual to recognise the harm that they have committed against another to make amends.

José Medina builds on Fricker's discussion by examining different ways to "remove epistemic obstacles and to achieve cognitive melioration not only for particular individuals and groups, but for the entire social fabric" (2013, p. 25). Medina makes a valuable contribution in unpacking the different communicative repertoires utilised by marginalised individuals who engage in epistemic resistance to challenge injustice. While the author discusses the credibility deficits that marginalised groups experience as speakers which contribute to testimonial injustice, he concretises how individuals resist censure and silences using different communicative means in the semiotic order, whether written, verbal or symbolic, to give voice to their ideas, thoughts and feelings and share them with others. Medina goes on to disrupt the notion of a monolithic public domain that *only* houses knowledge and resources produced by those in power. The author conceptualises multiple publics and hermeneutical resources embedded in the dominant order that are created by marginalised groups based on their social experience. While marginalised individuals may face epistemic disadvantages in being discredited as legitimate speakers (testimonial injustice), or they may have limited access to collective hermeneutical resources of the state, they also have epistemic advantages of their own that they can use to their benefit to resist subjugation.

Medina's point that different groups occupy, interact and compete in multiple publics can be compared to Sheller's point that marginalised groups, or "citizens from below," create counter-publics or spaces to resist domination and to affirm themselves and their lives (2012, p. 41). While Medina is well aware of the individual and systemic discrimination against underrepresented groups, he does not seek to locate their experiences in absolute victimhood. The author states that less privileged groups "use [their] epistemic resources and abilities to undermine and change oppressive normative structures and the complacent cognitive-affective functioning that sustains those structures" (Medina, 2013, p. 3). Medina also considers the heterogeneity of people's cognitive-affective responses in relation to power, oppression and resistance by situating their experiences within different social contexts. He states that:

> We are all affected by the epistemic obstacles and distortions that arise in situations of oppression. But differently situated groups and subjects have different epistemic predicaments: their epistemic deficits are different, and their resources to overcome these deficits and to resist dominant ideologies are also different. Epistemic oppression is not an equal opportunity institution: it affects all of us, but not all of us equally. (2013, p. 28)

Engaging in intersectional politics and coalition-building across different social movements is messy, and Medina sheds light on the different ways that knowledge is created, communicated and interpreted within, and among, groups that may promote or undermine justice and social change. Moreover, Medina embraces the chaos and the confusion that comes with participatory democracy, pluralism and the spontaneity of social movements because he sees possibilities in hermeneutical gaps when hegemonies, from inside and outside, are temporarily destabilised.

Drawing on Fricker's and Medina's works on epistemic injustice, in the next sections I will consider how the sexual rights of LGBTQ persons are compromised by epistemic injustices in the Commonwealth Caribbean. First, in terms of testimonial injustice, their right to self-identification and self-expression is compromised due to their not being perceived as credible speakers, which results in communicative faults between themselves and others around them. Second, in terms of hermeneutical injustice, the intimate life of sexual minorities related to partnerships, relationships, family and reproduction is obscured from collective understanding due to interpretive gaps in a heteropatriarchal social order. Moreover, I argue that the non-recognition of LGBTQ social experience by those in power compromises the creation of new knowledge and the pooling of hermeneutical resources that are needed to mitigate social injustices and to achieve a more democratic understanding of sexual citizenship.

Methodological Considerations

The methodology for this chapter comprises mixed data sources that include primary and secondary materials. To problematise the relationship between epistemic injustice and the impact of homophobia and transphobia on the lives of LGBTQ persons in the Commonwealth Caribbean, with special focus on Barbados, an interpretive analysis is used to review legislation and print media on the issue. I also incorporate the experiences of LGBTQ persons in Barbados based on a study[1] that was conducted in 2014–2015. Twenty-five LGBTQ persons, predominantly of African-Caribbean decent, shared their accounts in the study. The sample comprises 14 gay men of which four individuals identified as gender non-conforming or queer, eight lesbians of which four individuals identified as gender non-conforming or queer, two bisexual women, and one trans woman. Over two-thirds of the participants were in the age range of 18–30 years old. Qualitative methodology was employed in the form of semi-structured interviews to centre the experiences of the participants, which allowed the researcher to examine queer realities from multiple vantage points. Through a qualitative methodological approach, subjects are validated as situated knowers of their own realities and as those whose feelings, thoughts and actions have epistemic value.

To capture LGBTQ experiences of oppression and their resistance to prejudice and discrimination within their communities and social structures, questions covered the thematic areas of state power and legislation; discrimination, rights and equality; culture, social environment and visibility; family, socio-sexual unions and social networks; and resistance, activism and social justice. The data were coded through Dedoose (qualitative and mixed method data), and the responses were analysed through an interpretive analysis that considers power, knowledge and injustice through an interdisciplinary and intersectional lens based on decolonial, feminist and queer perspectives. Alias names were assigned to participants in the study to protect their anonymity.

Testimonial Injustice, Silences and LGBTQ Invisibility

So why do we need to speak our existence? Not all of us do, not all of us can. But some of us must speak, because we believe that silence will not protect us. We speak not to convince anyone of our existence, but to sing ourselves to ourselves and to remind those who despise us of our

[1] The Sexual Equality and Human Rights Youth project was conducted by co-investigators Charmaine Crawford and Shari Inniss-Grant, Institute for Gender and Development Studies, Nita Barrow Unit, the University of the West Indies, Cave Hill Campus, Barbados.

humanity—of our living, loving presence in the midst of their loathing. (King, 2008, p.194).

Rosamond King (2008), in her piece "More Notes on the Invisibility of Caribbean Lesbians," invokes Audre Lorde as she ponders the challenges that lesbians face with invisibility in their societies and the existential angst that silence brings with it in their daily lives. While King does not think that all silences should be broken, since some individuals are more vulnerable than others, she does, however, suggest that speaking up can be liberating for those who are able to do so. While personal integrity is strengthened in speaking one's truth in the face of oppression, testimonial injustice impedes a marginalised person's ability to do so. To counter this impairment, Medina considers the different ways that individuals may attempt to share their stories with others when they encounter opposition. He states that "testimonial exchanges are those in which communicators participate as knowers and possible epistemic benefits can be obtained" (2013, p. 28). For the author, these exchanges can be communicated through written, oral or creative modalities that are direct and indirect. Silence can also be a part of a testimonial repertoire. But testimonial exchanges are difficult for gay men and lesbians who have not disclosed their sexual orientation to their families and friends. Their reluctance in "coming out," if they desire to do so, is rooted in fears of being negatively judged or ostracised by their loved ones. Jody, who is in her 20 s and identifies as queer, recounts her journey of coming out to her family:

> I did have to come out to my parents at a time and they did not take it too well; my dad stopped talking to me for a while, my mother tried to get me healed, my brother though he asked me. He's been asking me if I was gay since I was like yay high and then one day we were, we went to vacation—he lives abroad—and we were going to a restaurant and we got out the car and he looks at me and he's just like, 'you're gay, huh, you're gay gay?'

In this testimonial exchange, Jody's brother did not give her a chance to disclose her sexuality to him on her own terms. Instead, he asked directly, which confirmed his suspicions. Jody's behaviour and silences throughout the years were deciphered by her brother who, thankfully, did not ridicule or reject her for being a lesbian. While Jody was supported by her brother, her parents, on the other hand, rejected her message as a speaker and knower of her own truth. Fricker notes that "the primary harm of (the central case of) testimonial injustice concerns exclusion from the pooling of knowledge owing to the identity prejudice on the part of the hearer" (2007, p. 162). Jody's testimonial exchange with her parents was difficult due to their homophobic attitude that devalued her and her experiences. In this case, testimonial injustice was operable in two ways: (1) Jody's father employed silence by refusing to speak to her when she initially discussed her sexuality. When he resumed speaking to Jody, he denounced homosexuality as abnormal, and he wanted their father–daughter relationship to continue without any mention of her lesbianism; and (2) Jody's mother rejected her message based on her religious beliefs that homosexuality is unnatural and a sin. She tried to spiritually counsel Jody onto the right path to overcome or cure her affliction—lesbianism. Jody's mother used religious moralism to dismiss her experience and shame her into thinking that lesbianism is an "unnatural" affliction that needs to be cured (Crawford, 2019).

Jody's mother's intolerance towards homosexuality is not surprising given the indoctrination of colonial Christianity in the region. Lazarus' (2015) discussion on the influence of Christianity in shaping post-independence nation-building efforts in the Commonwealth Caribbean is instructive here because she unpacks how a heteronormative Christian sexual citizenship has defined and legitimised what it means to be a man or woman in civil society through respectability politics. Homophobia produces epistemic harms of unbelonging that can malign the dignity of gay men and lesbians and keep them separated from others. Baker explains that dignity is the "the ability to establish a sense of self-worth and self-respect and to appreciate the respect of others" (2017, p. 134) but gay men and lesbians experience feelings of unworthiness due to homophobia. Thus, epistemic injustice undermines the dignity of gay men and lesbians because they do not feel valued and accepted by others, such as in Jody's case.

Guy, a gay man in his early 20 s, recounts his experience of not feeling respected by others because of his sexual orientation:

> I think because once again it's being a minority, you're not heard, you're usually misunderstood. And I think sometimes that can be challenging for people because at the end of the day, you're still a person, and you're still human. You still have feelings just like anyone else, and a voice, and a reason, and you have life, and we're all human. It's human nature to want to be heard, and understood, and loved, and cared for ….

Based on Guy's account, he does not feel valued when he is not listened to or when he is negatively judged because of his sexuality. Guy makes the connection between being "not heard" and being "misunderstood," signalling how epistemic harms can occur because of prejudice that invalidates the identity and knowledge of the speaker, resulting in miscommunication and misinterpretation (hermeneutical faults) between differently situated individuals. This goes back to Fricker's point that when it comes to epistemic injustice, "identity power" influences *who speaks for whom* or *who listens to whom* based on hierarchal relations within the social order. These dynamics impact the willingness and ability of marginalised individuals to speak out against injustice. When individuals who are disadvantaged decide to articulate their pain, they do so, as Lorde notes, "at the risk of having it bruised or misunderstood" (2017, p. 1). Guy's longing to be "understood, loved and cared for" goes back to Medina's (2013) point about *empathy* being an important principle in social justice activism. To show compassion towards an individual with a different experience or perspective than your own may be helpful in bridging gaps across difference and in mediating and ameliorating conflict that might arise in social interactions.

Because homosexuality is perceived as deviant, unnatural and morally corrupting within heteropatriarchal logic (Alexander, 1994; Crawford, 2019; Robinson, 2009), sexual minorities are othered and marginalised as sexual citizens in the Commonwealth Caribbean. While heterosexuals do not have to disavow their sexual identity to be accepted by others due to the privilege that they hold in society, Guy, and other gay men and lesbians, must fight for their "humanness" to be recognised and respected. Not surprisingly, many LGBTQ individuals tend to embrace human rights discourse to assert their inalienable rights as persons to counter prejudice and discrimination (Murray, 2006). While human rights discourse may be strategically used to influence public policy and the state interventions in establishing minimum standards in protecting the civil rights of minority groups based on universalist principles that recognise our shared humanity, it does effectively overturn the systems and hegemonies that produced and continue to produce social inequalities, in the first place, because it aligns with an integrationist approach that addresses "differences" within a dominant group or social order through inclusion and tolerance.

The inability or unwillingness to accept gay men and lesbians for who they are, instead of simply tolerating them, means that their credibility is constantly under attack because knowledge about them has not been fully accepted and integrated into society in a meaningful way. For instance, Brian, a gay man in his 40 s, expresses that homophobia makes him feel like he is not a "complete or full person." He goes on to state that he feels that gay men are treated as "suspicious characters" who are up to no good compared to heterosexual men. Brian's account reflects how colonial heteropatriarchal masculinist ideals are embedded in dominant notions of masculinity in the post-independent era that rely on demonising and subjugating black homosexual men through discriminatory anti-buggery laws and harassment by law-enforcement officers. Black and brown gay men are perceived as disorderly and dangerous to the integrity of hegemonic (heterosexual) masculinity (Lewis, 2003), so they are policed and criminalised as suspicious characters or non-credible subjects. Gay men in the Commonwealth Caribbean have reported that they are not taken seriously by police officers when they are victims of homophobic violence and intimate partner violence (Human Rights Watch, 2017). The devaluation of gay men's existence does not only occur in the state and law but also in popular culture whereby homophobic sentiments have been directed towards gay men. Larcher and Robinson (2009) discuss their activism in "Stop Murder Music," an international campaign to protest homophobic lyrics in dancehall music in the Caribbean and wider diaspora that have condoned violence and the killing of gay men.

Gay men experience testimonial injustice when they are silenced or ignored altogether when they speak out against victimisation, and they fail to receive equal protection under the law. Alex, a gay man in his late 20 s, recalls his traumatic experience in seeking assistance from the police after he was a victim of sexual assault. When he reported the incident to police officers, they laughed at him. Not only did Alex have to deal with the indignity of being discredited as a victim of sexual assault as a gay man, but he also had to accept the possibility that nothing would be done about his complaint. Alex felt demoralised and fearful after the violent incident, and the lack of compassion and accountability shown by the police exacerbated his anxieties about his personal safety. Due to his experience of testimonial injustice, Alex aptly states, "that's why some men wouldn't even go the police."

Medina (2013) notes that since epistemic oppression is situated for individuals, society must be democratically transformed so that a variety of viewpoints are heard and validated in laws, policies and practices to mitigate injustice. But listening to someone speak does not mean that they will be understood if there is a scarcity in hermeneutical resources. Medina goes on to explain how testimonial injustice and hermeneutical injustice interplay with one another:

> Because of difficulties in expressing and interpreting certain things—because of hermeneutical insensitivities —people's credibility can get undermined; but also their lack of credibility can call into question the intelligibility of their formulations and interpretations, especially when they are advancing new meanings and struggling to make sense in the face of widespread hermeneutical limitations. (2013, p. 96)

In Alex's case, both testimonial injustice and hermeneutical injustices occurred when he reported that he was sexually assaulted. First, in terms of testimonial injustice, Alex, as a gay man, lacked credibility as a subject of sexual assault because he did not fit the image of prototypical victim—a cisgender heterosexual woman. Haynes and DeShong note the negative way the media cover transgender and gender non-conforming persons when it comes to violence, commenting that "the media not only uses sensationalist language to describe acts of violence against transgender and gender non-conforming persons, but also uses language which trivializes the violence which they are reporting" (2017, p. 115). Second, there are interpretive deficits in laws, policies and practices that do not fully capture and address violence against LGBTQ persons in both private and public domains. Due to structural inequality born of heterosexism and homophobia, the police, as an entity of the state, demonstrated hermeneutical insensitivities through their indifference to Alex's complaint of sexual victimisation as a gay man.

Given the pervasiveness of violence against women in society, sexual violence (such as in the case of rape) has commonly been investigated by feminists and social advocates through a male perpetrator and female victim dyad. While this is justifiable to combat gender-based violence against women and to challenge men's heteropatriarchal claims over women's bodies, it is also important that collective hermeneutical resources are aggregated to better understand how violence occurs beyond a cisgender heterosexual model. This is necessary in order to combat violence and provide legal redress for survivors of abuse. Although domestic violence legislation in the Commonwealth Caribbean is gender neutral in that it applies to both male and female victims (Robinson, 2000), it is nonetheless interpreted through a heteronormative lens whereby spouse is defined as husband and wife through marriage or co-habitation.

This myopic heteronormative trope in the law invariably compromises the right to equal protection before the law for sexual minorities when it comes to domestic violence. For example, the Barbados' Domestic Violence (Protection Orders) Amendment Act XE "Barbados' Domestic Violence (Protection Orders) Amendment Act", 2016 provides a stronger mandate for police to respond to domestic violence and to execute emergency protection orders, and provides a better definition of domestic violence and covers persons in different domestic relationships (Browne & Nicholls, 2016; UN Women Caribbean Portal, 2022). The Act states that "'domestic relationship' means the relationship between a perpetrator of domestic violence and victim who is a spouse, former spouse, child, dependant or other person who is considered to be a relative of the perpetrator by virtue of consanguinity or affinity and includes cohabitational and visiting relationships" (Domestic Violence Amendment Act, 2016). While the domestic violence amendment committee, comprising governmental, NGO and civil society representatives, drafted the recommendations for amendment to be inclusive of all domestic relationships, including same-sex relationships, this was not the result. When the Bill was debated in parliament a politician threatened to vote against the amendment if it included same-sex domestic partnerships and, not surprisingly, other conservative dissenters supported his position (Mohammed, 2017). Browne and Nicholls rightly argue that "one flaw of the Act is it limits a cohabitational relationship to one based on the parties living as husband and wife, which excludes people who live-in homosexual relationships" (2016, p. 4).

In addition to this, transgender and gender non-conforming individuals encounter both gender and sexual stereotypes in their lives. As a butch lesbian, Jody articulates her frustration in being negatively judged because of gender identity and sexuality:

> I just got frustrated and said listen, I can't live a life that I'm not. You need to understand that this is me and stop asking me: You don't find this boy nice? Wear this dress and that .... And I was like, okay, look, no!

Jody's testimonial exchange with her family reveals their attempt to silence and delegitimise her as a credible speaker because of her identity. The testimonial insensitivities Jody endures cannot be separated from the hermeneutical insensitivities that inform the interpretive prejudices about her as a woman. This is evident through the moral authority shown by family members who harassed Jody about her non-feminine attire and about not having a boyfriend. The interpretive deficits that invalidated Jody's social experience reflect the normalisation of Enlightenment heteropatriarchal ideologies in public consciousness that have come to define femininity and masculinity in Caribbean modernity. Alexander cautions us about the pervasiveness of hierarchies of power in advanced capitalism that affect us as sexual citizens. She is troubled by a "sexualized hegemony" that is "simultaneously knitted into the interstices of multiple institutions as well as into everyday life" (2005, p. 4). So, while lesbians can exercise personal agency in celebrating their love of themselves and other women in their personal lives, their identities and relationships continue to be stymied by respectability politics in society as seen through the moralistic ideals that have come to define "womanhood" based on heteronormativity and black middle-class gender ideology in the Commonwealth Caribbean.

Hermeneutical Injustice and LGBTQ Marginalisation

Medina (2013) notes that there is a reciprocal relationship between social inequality and epistemic injustice because both derail fairness in the production of knowledge and sharing among individuals. The author goes on to argue that "in a situation of oppression, epistemic relations are screwed up. Inequality is the enemy of knowledge: it handicaps our capacity to know and to learn from each other" (2013, p. 27). Thus, the epistemic injustices that Caribbean gay men and lesbians encounter due to heterosexism and homophobia are multifaceted. Fricker notes that hermeneutical injustice takes place on a personal level "when a gap in interpretive collective resources puts someone at an unfair advantage when it comes to making sense of their social experiences" (2007, p. 1). Since heterosexism—the primacy, legitimacy and institutionalisation of opposite sex sexual relations—is over-determined in social structures, hermeneutical injustice can take place whereby queer knowledge and experiences are blocked from collective understanding due to prejudice and discrimination. Gay men and lesbians are viewed as inferior, and they are "othered" as sexual citizens due to their intimate relations being seen as promiscuous, primal and unnatural. Since heterosexuality is privileged and normalised in society—so much so that the symbols, rituals and practices associated with opposite-sex intimacies are interwoven into private-public life through the family, state, law and popular culture—its recognition and acceptance is firmly established in our collective understanding and imagination. Conversely, hermeneutical faults born of heterosexism, homophobia and transphobia obscure the gamut of queer relationships both sexual and non-sexual from collective understanding and public view.

Morgan's situation as a lesbian in her 30 s is a good example of how homophobia produces hermeneutical disadvantages both on the individual and structural level for gay men and lesbians. She recalls her mother, Ms. Smith, not only rejecting her as a lesbian but attempting to remove her from the household when she brought a female friend home:

> The [police] let her know that they cannot remove anyone for being who they are. They can't remove anyone for their sexual orientation. The only reason they can remove this person, is if the person is being disrespectful to your place or cause you harm. And the police just advise me that this is mummy's castle, and whatever she says goes. Do not bring anyone of that 'nature' to the house, and around her house to make her feel uncomfortable. If I want to do anything, do it away, because that's who I am, that's my sexual orientation and I still have my life to live but do it away from mummy's house.

Morgan's case demonstrates some of the challenges that LGBTQ young adults face while living at home and being financially dependent on their parent(s). They find it difficult to confidently share their feelings about their identity, romantic partners and life aspirations. Dixon-Mueller et al. (2009) note that an ethics of sexual rights do not only include the recognition of one's sexual identity and the freedom to engage in consensual pleasurable sex, but they also include the freedom to choose sexual partners free from coercion. But gay men and lesbians face restrictions when it comes to their sexual rights because sexual citizenship is hermeneutically conceived and normalised through a heteronormative logic. Thus, Morgan's right to self-determination and ability to freely choose her sexual partner was undermined by her mother's prejudice against LGBTQ people that clouded her interpretive abilities in understanding her daughter's situation and in caring about her well-being. Morgan tried to exercise some agency by not living a closeted life, but her actions were met with disapproval from her mother and community members who gossiped about her "lifestyle." With hermeneutical injustice on an individual level, unfairness can take place when an egregious act is committed against an individual because of their identity but also occurs when the oppressor fails to acknowledge the harm that they have caused to the other person, and they do not seek to change their behaviour (Fricker, 2007; Medina, 2013). Even though Ms. Smith's homophobic rant humiliated Morgan, she felt justified in sanctioning her daughter for what she believed to be morally corrupting behaviour (Crawford, 2019). This resulted in her attempting to punish Morgan for her disobedience by calling the police to remove her from the household.

Richardson states that the "displacement of the homosexual as other reifies the moral superiority of heterosexuals in upkeeping a just and descent society" (2000, p. 102). Ms. Smith could not overcome her hermeneutical liabilities because she saw Morgan as an undesirable homosexual. Her prejudice and lack of empathy shown towards Morgan kept her separated from her daughter. Ms. Smith's behaviour reflects what Medina refers to as epistemic arrogance, which is "when the subject becomes absolutely incapable of acknowledging any mistake or limitation, indulging in a delusional cognitive omnipotence that prevents [them] from learning from others and improving" (Medina, 2013, p. 29). Heterosexism is over-determined in culture and social structures with its own hegemonic repertoire. Thus, Ms. Smith felt emboldened to utilise the arm of the state (police) to remove her daughter from the household, even if it meant leaving her child homeless. Although the police had no grounds to remove Morgan as an occupant of the household simply because of her sexual orientation, they nonetheless reminded her about the deviance associated with homosexuality by advising her not to bring people of a certain "nature" into her mother's house.

Morgan's situation sheds light on how queer individuals must constantly negotiate visibility and invisibility in their families and communities. There is a dialectical process between "coming out" or the disclosure of one's sexuality and/or identity and concealment of one's sexual/gender identity (or being closeted) that is of epistemic importance. Medina notes that marginalised individuals based on their "different epistemic predicaments" deal with oppression in a variety of ways that can be either hermeneutically advantageous or disadvantageous to them (2013, p. 98). So, "coming out" is not one size fits all or a meta-experience for sexual minorities. The *epistemology of coming out*, or alternatively the *epistemology of the closet* as coined by Sedgwick (1990), is a subjective and iterative process that involves persons strategically disclosing their same-sex desire or homosexual identity within a heteronormative order. Thus, coming out is a fluid process that varies temporally depending on a person's intrapsychic development, personal circumstances and social background. For example, queer existence is mediated through intersectional identities and oppressions whereby affluent white gay individuals have race and class privileges to buffer the effects of discrimination compared to economically marginalised black gay men and lesbians whose sexuality is viewed as deviant through the prism of heterosexism and racism. Hill Collins argues that "the core binary of normal/deviant becomes ground zero for justifying racism and heterosexism" (2005, p. 97). Black LGBTQ persons may have fewer epistemic advantages to lessen or mitigate injustice, contributing to hermeneutical marginalisation due to the lack of support and resources within their communities and wider society to overcome multiple oppressive conditions. Medina notes that "when it comes to hermeneutical gaps, it is crucial to pay attention to the communicative processes in which subjects struggle to make sense to themselves of what they cannot yet communicate to others, especially to those others who do not share their predicament" (2013, p. 98). Homophobia produces its own interpretive vulnerabilities for Caribbean LGBTQ persons, resulting in them being less willing to share information about themselves with others to protect themselves against ridicule, ostracism or violence.

Overcoming hermeneutical gaps may take some time for individuals who are coming to terms with their sexual and gender identity within a cisgender heteronormative society. Riley, a lesbian in her 20 s, shares her experience of exploring her sexuality from adolescence into adulthood. Riley considered herself a tomboy growing up and she did not view her gender non-conforming behaviour as a problem at that time. As a teenager, she had romantic feelings for other girls and had her first kiss with a girl at 14 years old. In her late teens, Riley became romantically involved with a young woman who she liked but the relationship did not last long because her parents were growing suspicious of her *peculiar* (read: queer) behaviour because she did not have a boyfriend, and her appearance was not stereotypically feminine. Riley experienced hermeneutical marginalisation as she struggled to make sense of her situation. On the one hand, she did not want to disappoint or become ostracised from her family by coming out as a lesbian, but on the other hand, she was attracted to women. Whether it was due to familial and/or societal pressure, Riley married a man and had a child as a young adult. She divorced her husband several years later and then resumed dating women. Riley confessed that she was "living a double life" while married because she still had feelings for women, and she occasionally flirted with her female friends.

Riley's situation reflects the messiness of hermeneutical gaps for sexual minorities who must navigate their same-sex or bisexual desire during the maturation process amidst a heteronormative environment. It takes a lot of hard work to overcome hermeneutical vulnerabilities for marginalised individuals because oppressive situations are traumatic, which makes it difficult to work through painful memories and feelings and communicate them to others. While the interpretive gaps that Riley encountered in coming to terms to with her sexuality align with representations of sexual fluidity in the region related to bisexuality or individuals who have sex or are sexually attracted to both genders without claiming a sexual identity, such as men who have sex with men (MSM) or women who have sex with women (WSW), her story offers a more nuanced vantage point about the heterogeneity of queer articulations that cannot be easily classified into mutually exclusive sexual categories in the Caribbean. Gosine (2009) notes that male bisexuality is seen as a problem in the region by contributing to the spread of HIV/AIDS and in threatening the heterosexual nuclear family and social fabric of society, so it has been managed through public health initiatives and a non-identification with being gay, such as MSM.

Riley's sexual journey was an iterative process occurring over many years that was cumbersome, involving different ways of being both socially and sexually until she was able to accept herself and her sexual preferences confidently and comfortably. Coming to terms with one's sexual identity, and by extension the coming out process, is not easy and can be filled with confusion, angst and miscommunication, producing hermeneutical insensitivities and hurtful encounters with others. Although Riley was able to improve her "hermeneutical sensibilities" over time, she had to make difficult decisions about her life that required mental and emotional maturity.

While homosexuality in the Caribbean is often reduced to keeping same-sex relations within the private sphere away from public view, Morgan's and Riley's experiences demonstrate that when it comes to sexual citizenship the intimate and familial traverses both private and public domains. The erotic autonomy of Caribbean LGBTQ youth is often constrained by their inability to exercise their sexual rights due to heteronormative standards that reproduce social inequalities that devalue their social experience from general comprehension (Alexander, 1997). While some

LGBTQ youth find outlets to explore their sexuality and authentically express themselves with their peers, such as social clubs and sports at school, many others encounter emotional and mental trauma in having to deal with the process of sexual maturation, exploration and intimacy in hostile situations and environments. They, in turn, experience hermeneutical marginalisation because they lack support and the interpretive resources to challenge injustices. LGBTQ youth experience higher rates of poverty, homelessness and mental illness compared to other youth populations due to the homophobia, transphobia and ostracism that they face from their families.

In conclusion, in this chapter I explored how epistemic injustices operate on an individual level to invalidate the social experiences of LGBTQ persons in the Commonwealth Caribbean. Testimonial injustice due to prejudicial attitudes towards LGBTQ and gender non-conforming persons produces cognitive-affective deficits between LGBTQ persons and others whereby they are not heard as credible speakers. This negatively impacts their self-esteem and jeopardises their ability and willingness to communicate with their family and friends and may impede their ability to be treated fairly in the eyes of the law. Testimonial injustice is also linked to hermeneutical injustice on the individual level. When LGBTQ individuals are not valued or validated a speakers and knowledge producers, they may internalise their oppression. This results in hermeneutical marginalisation whereby they are able to tap into inner reserves and pool available resources to counter their oppression.

REFERENCES

Alexander, J. M. (1994). Not just (any)*body* can be a citizen: The politics of law, sexuality and post-coloniality in Trinidad and Tobago and Bahamas. *Feminist Review, 48,* 5–23.

Alexander, J. M. (1997). Erotic autonomy as a politics of decolonization; an anatomy of feminist and state practice in The Bahamas tourist economy. In M. Jacqui Alexander & C. Mohanty (Eds.), *Feminist genealogies, colonial legacies, democratic futures* (pp. 63–100). Routledge.

Alexander, J. M. (2005). *Pedagogies of crossing: Meditations on feminism, sexual politics, memory, and the sacred.* Duke University Press.

Baker, S. J. (2017). Is it safe to bring myself to work: Understanding LGBTQ experiences of workplace dignity. *Canadian Journal of Administrative Sciences, 34,* 133–148.

Browne, F., & Nicholls, A. (2016). *Domestic violence: Victim protection and intervention.* https://barbadosunderground.files.wordpress.com/2016/02/domesticviolence_alicianichollandfeliciabrowne.pdf

Crawford, C. (2019). Unbearable knowledge: Sexual citizenship, homophobia and the taxonomy of ignorance. *Journal of Eastern Caribbean Studies, 44*(2), 115–144.

Dixon-Mueller, R., Germain, A., Fredrick, B., & Bourned, K. (2009). Towards a sexual ethics of rights and responsibilities. *Reproductive Health Matters, 17*(33), 111–119.

Domestic Violence (Protection Orders) (Amendment) Bill. (2016). (Barbados). https://www.barbadosparliament.com/uploads/bill_resolution/907d022cc7 6d0c58b3353e80836ba3e6.pdf

Fricker, M. (2007). *Epistemic injustice power & ethics of knowing.* Oxford University Press, Inc..

Gosine, A. (2009). The Heteronationalism of MSM. In C. Barrow, M. de Bruin, & R. Carr (Eds.), *Sexuality, social exclusion and human rights: Vulnerability in the context of HIV* (pp. 95–115). Ian Randle Publishers.

Haynes, T., & DeShong, H. (2017). Queering feminist approaches to gender-based violence in the anglophone Caribbean. *Social and Economic Studies, 66*(1 & 2), 105–131.

Hill Collins, P. (2005). *Black sexual politics: African Americans, gender, and the new racism.* Routledge.

Human Rights Watch. (2017). *I have to leave me behind: Discriminatory Laws against LGBT people in the eastern Caribbean* report. United States. https://www.hrw.org/report/2018/03/21/i-have-leave-be-me/discriminatory-laws-against-lgbt-people-eastern-caribbean

King, R. (2008). More notes on the invisibility of Caribbean lesbians. In T. Glave (Ed.), *Our Caribbean: A gathering of gay and lesbian writing from the Antilles* (pp. 191–196). Duke University Press.

Larcher, A. A., & Robinson, C. (2009). Fighting murder music: Activist reflections. *Caribbean Review of Gender Studies, 3*, 1–11.

Lazarus, L. (2015). Sexual citizenship and conservative Christian mobilisation in Jamaica. *Journal of Eastern Caribbean Studies, 40*(1), 109–140.

Lewis, L. (2003). *The culture of gender and sexuality in the Caribbean* (pp. 94–125). University Press of Florida.

Lorde, A. (2017). The transformation of silence into language and action. In *A. Lorde your silence will not protect you, anthology* (pp. 1–6). Silver Press.

Medina, J. (2013). *The epistemology of resistance: Gender and racial oppression, epistemic injustice, and the social imagination.* Oxford University Press.

Mohammed, R. (2017). *B-GLAD voices: The LGBT country report.* Barbados, Gays, Lesbians and All-Sexuals against Discrimination. https://www.academia.edu/36091654/VOICES_-_Barbados_LGBT_Country_Report

Mohanty, C. T., & Carty, L. (Eds.). (2018). *Feminist freedom warriors: Genealogies, justice*. Haymaker Books.

Murray, D. (2006). Who's right? Human rights, sexual rights and social change in Barbados. *Culture, Health & Sexuality, 8*(3), 267–281.

Richardson, D. (2000). *Rethinking sexuality*. Sage.

Robinson, T. (2000). Fictions of citizenship, bodies without sex: The production and effacement of gender in law. *Small Axe, 7*, 1–27.

Robinson, T. (2009). Authorized sex: Same-sex sexuality and law in the caribbean. In C. Barrow, M. de Bruin, & R. Carr (Eds.), *Sexuality, social exclusion and human rights: Vulnerability in the context of HIV* (pp. 3–22). Ian Randle Publishers.

Sedgwick, E (1990). *Epistemology of the Closet*. University of California Press.

Sheller, M. (2012). *Citizenship from below: Erotic agency and Caribbean freedom*. Duke University Press.

UN Women, Caribbean Portal. (2022). *Overview of country gender equality status (Barbados)*. https://caribbean.unwomen.org/en/caribbean-gender-portal/barbados

The Precarity of Sexual Citizenship: Hermeneutical Injustice, the Law and LGBTQ Rights

Abstract In this chapter, I explore how heterosexism, homophobia and transphobia operate on a structural level in state, law and institutions to disadvantage LGBTQ persons and delegitimise them as sexual citizens when it comes to their identity, sexual conduct and socio-sexual relations with others. I consider how coloniality and heteropatriarchal ideologies in anti-buggery and vagrancy laws have normalised the outsider status of LGBTQ persons through structural hermeneutical injustice. This goes beyond personal biases that sometimes get rationalised as ignorance or occasional occurrences. Instead, it is operationalised within institutions and by those in power. The test to Caribbean constitutionalism in reconciling colonial and post-colonial imaginings of sexual citizenship, through strategic litigation in decriminalisation cases as it relates to sexual rights for LGBTQ person, is the focal point of this chapter. Finally, I discuss how the shifts in Caribbean jurisprudence when it comes to LGBTQ rights are linked to both internal and external factors. But more importantly, I argue that real social change in this area could not have been actualised without the leadership, grassroots organising and epistemic resistance of LGBTQ activists and groups in the region and the Caribbean diaspora in the twenty-first century.

C. Crawford, *Gender, Sexual Citizenship and Epistemic Injustice in the Caribbean*, https://doi.org/10.1007/978-3-031-83493-6_4

Keywords Structural hermeneutical injustice • Strategic litigation and anti-discrimination • Caribbean Constitutionalism • Caribbean Court of Justice • Decriminalisation cases: homosexuality and transgenderism • Religious backlash • Counter-publics and epistemic resistance • Caribbean LBGTQ organising • Caribbean LGBTQ rights and sexual citizenship

> We should be free to define our own identity and concept of life and self – including our sexuality—without the compulsion of the state. When the state insists, through law or otherwise, that we must be sexed in a specific way—be it heterosexual, monogamous, nuclear family-oriented, married, or mother— it strikes at the heart of dignity as human beings, and treats us as unworthy as persons *and as* citizens. (Robinson, 2009, p. 3)

Robinson (2009) recognises that sexuality is an integral part of being human, but it is not neutrally constructed when it comes to state power. Caribbean states, as constitutional democracies, have replicated oppressive, antiquated colonial laws and practices in their modern nation-building efforts, particularly against certain members of their citizenry, in the quest for independence and sovereignty. Since "heterosexual reproduction was a preoccupation of the national project of the Caribbean" (Robinson, 2009, p. 7), LGBTQ persons find it difficult to exercise their sexual rights due to discriminatory laws that regulate sexual conduct by criminalising same-sex relations and not legally recognising same-sex partnerships. Educated, middle-class black and brown elite men took power and led Caribbean countries to independence through reformist politics derived from Enlightenment bourgeois heteropatriarchal constructions of citizenship and nationhood. Alexander (2005, p. 4) aptly states, "I am concerned with the multiple operations of power, of gendered and sexualised power that is simultaneously raced and classed yet not practised within hermetically sealed or epistemically partial borders of the nation-state." Despite this, tensions between the state, leaders and LGBT activists create the hermeneutical friction necessary for deconstructing and decolonising laws and practices that have compromised the rights and dignity of LGBTQ persons. Through robust civil society engagement, opportunities emerge to create counter-narratives and counter-publics to mitigate social injustices.

In this chapter, I will focus on how hermeneutical injustices are reproduced in law and state to disadvantage Caribbean LGBTQ persons. First, I will consider how LGBTQ activism is thriving in the region and how groups create counter-narratives through personal and collective resistance to challenge stigma and discrimination. Epistemic resistance, in particular, helps empower and elevate the consciousness of LGBTQ youth through community-building and mobilisation efforts. Second, I will critique the role of political leaders in not fully protecting the rights of sexual minorities and in fostering an environment of intolerance towards them. Finally, I will examine the critical role that strategic litigation has played in decriminalisation cases, particularly in overturning prohibitive buggery (anal sex) and transgender laws. Anti-racist, decolonial, feminist and queer analyses are beneficial to legal hermeneutics in fostering epistemic justice, aiming to upend discriminatory legislation and socially transform society for the betterment of all.

CONFRONTING HEGEMONY: COUNTER-PUBLICS AND LGBTQ EPISTEMIC RESISTANCE

Over the last 20 years, Caribbean LGBTQ persons and groups have initiated multiple forms of epistemic resistance against injustice, differentiated across age, class, race/ethnic and national lines, making them locally situated developments of grassroots queer consciousness in the region. Medina (2013, p. 3) notes that epistemic resistance entails acquiring and operationalising different "epistemic resources and abilities" to proactively combat the ideological and material basis of the dominant order. Caribbean LGBTQ activists and groups have demonstrated epistemic resistance by strategising with various actors, including feminists, lawyers, academics and national, regional and international NGOs, in fighting for sexual rights. Murray (2006) notes that various public forums in Barbados in the early 2000s dealt with gender and sexual rights. These efforts were "presented through different conceptual frameworks including, but not limited to those that are physiological, legal, moral, ethical, psychological, cultural, educational, or developmental" (Murray, 2006, p. 268). LGBTQ strategising included personal acts of resistance, with individuals drawing on the support of their family, friends and communities to care for and protect them, as well as collective mobilisation that politicised and publicised grievances to bring about change. Sheller's (2012) conceptualisation of sexual citizenship from below and the creation of "counter-publics" is indicative of epistemic resistance. She states that:

> Sexual citizenship is not just about personal rights or individual empower-ment, nor is it simply about state recognition of certain kinds of privacy, although it includes all of these. It is also concerned with collective processes, public spaces and forms of interrelatedness that are sexual and sexualized. Thus, it is a politics that is closely linked to public performances and perfor-mativity of public action and, hence, a politics in which bodies are central and cannot be ignored or bracketed as in classical liberal theory. (Sheller, 2012, p. 41)

This standpoint validates a variety of embodied sexual subjectivities engaged in individual and collective actions that are politically and epis-temically valuable in challenging the status quo in the public domain. This liberatory perspective moves beyond characterising the Caribbean queer experience as primarily private sexual acts or sexual praxis (Kempadoo, 2009) based on non-identification. Instead, it encompasses broader Caribbean queer social movements that collectively voice grievances to effect change. While these movements are organically formed, decentral-ised and may appear messy at times in our globalised time-space, their non-homogeneous formation reflects diverse, entangled local and diasporic partnerships within post-colonial relations.

Attai (2017) rightly critiques the homonationalist practices of some white Canadian organisations (and others) that have promoted a human rights mandate, with the help of enthusiastic gay Caribbean expatriates, to save and liberate poor Caribbean gay men from victimhood. Attai (2017, p. 101) states that this perspective "avails the Caribbean queer as a site through which imperialistic control is mapped onto bodies and communi-ties, notwithstanding the fact that such sexual praxes are indeed sites of resistance against prevailing control mechanisms." Thus, Caribbean LGBTQ activists use their lived experiences to challenge homophobia and transphobia in their communities and through broader social mobilisation.

Transgender women have courageously fought to carve out spaces in their communities, supporting queer folk and establishing counter-publics as drag performers in the tourist industry in Barbados (Murray, 2009). Chase, a transgender woman, points out that the visibility of transgender women "didn't emerge out of nowhere" but was sparked by the activism of individuals during the height of the HIV/AIDS crisis in the 1980s with the formation of United Gays and Lesbians against AIDS in Barbados (UGLAAB). Building on past queer activism in Barbados, in 2012, the Barbados Gays, Lesbians and All-Sexuals against Discrimination (B-GLAD) was formed by Donnya Piggott and Ro-Ann Mohammed. This group vocally advocated for LGBTQ rights in Barbados, critiquing the government's duplicitous stance on gay rights, denouncing violence against gay and transgender persons, and organising activities for queer youth to congregate and socialise with each other (Mohammed, 2017).

The heightened visibility of Caribbean LGBTQ groups in the public domain produces a catch-22 scenario. On the one hand, they use their platform to denounce discrimination publicly, but on the other hand, they are vulnerable to backlash in popular culture, media and the state. For example, on 23 June 2013, a press release appeared in *Nation News* about the formation of B-GLAD and its activities. Then, a week or so later, a counter-story was published in a newspaper with the title "Beware of BGLAD's offer of Help," which accused the group of being influenced by foreigners and seeking to taint the Christian values of Barbadian society to support their lifestyle and, much worse, same-sex marriage (Amlak, 2013). The author concluded that the two leaders should focus on "changing their lifestyle, which will bring them damnation, and not attempt to rally an entire nation to their follies" (Amlak, 2013, p. x).

The heterogeneity of Caribbean LGBTQ social movements has capitalised on national, regional and international partnerships, raising their visibility and creating a collective hermeneutical resource bank to express grievances and support one another. This aligns with Medina's point that "the aim of an epistemology of resistance here is to elucidate the practices and attitudes that can improve the social imagination and facilitate our openness to differences" (2013, p. 266). National[1] and regional[2] LGBTQ

---

[1] Jamaican Forum for Lesbians, All-Sexuals and Gays (J-FLAG), United Gays and Lesbians against AIDS Barbados (UGLAAB), Barbados Lesbians and All-Sexuals Against Discrimination (B-GLAD), Coalition Advocating for Sexual Orientation (CAISO)-Trinidad, Society Against Sexual Orientation Discrimination (SASOD)-Guyana, and Bahamas LGBT Equality Advocated.

[2] The Caribbean Forum for Liberation and Acceptance of Genders and Sexualities is an umbrella organisation that supports LGBTQ groups with their initiatives across the region. Countries that are served include St Lucia, Jamaica, Trinidad and Tobago, the Dominican Republic, Belize, Grenada, Guyana and Suriname. The Eastern Caribbean Alliance for Diversity and Equality (ECADE) is an umbrella organisation that serves the Eastern

organisations have created different counter-publics or spaces of dissident engagement and action, with members risking their lives in the face of homophobic and transphobic violence, to meet the needs of gay men, lesbians and transgender individuals, which is a crucial component of sexual citizenship.

For instance, Jamaica Forum for Lesbians, All-Sexuals and Gays (J-FLAG), Jamaica's flagship LGBTQ organisation, was founded in 1998 by a team of queer men and women. The organisation's commitment to advocating for the lives and rights of LGBTQ persons in Jamaica encompasses a wide range of human rights and social justice initiatives, including community services, social events, research, media content and policy advocacy. Organisers must weigh the cost between visibility and safety depending on their level of activism. Some LGBTQ activists, especially vocal leaders, have been harassed, victimised and even killed for speaking out, leading others to seek asylum abroad to protect themselves (Bowcott & Wolfe-Robinson, 2012; Lavers, 2014). However, those on the ground have found creative ways to engage in coalition-building, generate new knowledge, and pool resources and time towards public education, policy initiatives, and conventional and social media to raise awareness around LGBTQ rights.

In addition, intersectional politics are an integral part of queer organising, which has sparked meaningful connections with a new generation of feminists. These feminists interrogate the intersections of gender and sexuality through their teaching, research and activism at the Institute for Gender and Development Studies at the Mona, St Augustine and Cave Hill campuses of the University of the West Indies. Tech feminists use digital technology via online magazines, blogs and social media (such as Facebook and Twitter) to raise public awareness, protest and launch campaigns. This creation of cyber feminist counter-publics forms a vast epistemic toolkit of new and shared knowledge, resources and communities that have been valuable to feminists and queer activists alike. Haynes (2016, p. 3) states that "through practices of media creation, curating, reblogging, retweeting, sharing, and commenting across multiple social media platforms, Caribbean feminists knit together online communities that are often linked to on-the-ground organising and action."

Caribbean with training and technical support. Regional support extends to Antigua and Barbuda, Barbados, Dominica, Grenada, St Lucia, St Kitts and Nevis, St Vincent and the Grenadines, and Saint Martin.

The International Resource Network has been a critical LGBTQ knowledge bank since 2009. It gathers and disseminates news, materials and resources about Caribbean sexualities, actively connecting Caribbean queer activists and communities in the region and diaspora. In 2012, the International Resource Network published one of the first digital collections on Caribbean sexualities, showcasing scholarly and creative essays and writings co-edited by Rosamond S. King and Angelique Nixon with technical support from Vidyaratha Kissoon.

The capacity, services and resources of Caribbean LGBTQ organisations vary depending on their funding sources and programmatic needs. Financial uncertainty is a constant challenge for civil society organisations, especially in the global South, due to their reliance on internal and external donors. Some NGOs and state support tend to focus on public health services geared towards HIV/AIDS prevention in so-called high-risk populations (such as men who have sex with men [MSM] and sex workers), which inadvertently pathologises non-conventional sexualities (Gosine, 2009).

LGBTQ groups also seek grant funding from regional and international NGOs and country overseas offices (such as the US Embassy, Canadian High Commission, British High Commission and European Union) for capacity building, public education and outreach, youth training, and democratisation efforts through promoting inclusive public policy and legislation. The strength and visibility of Caribbean LGBTQ organising over the last couple of decades have been bolstered by grassroots organising and increasing public awareness of LGBTQ issues through media campaigns via the print press, television and social media.

Initiatives include, but are not limited to, (a) workshops on self-help, sexual health, reproductive health and well-being; (b) hosting and participation in social, cultural and political events such as parties, carnivals, Pride parades and protests; (c) leadership training and capacity building for LGBTQ youth; (d) partnerships with regional and international human rights NGOs on public policy, programmes and funding; (e) civil society engagement with the state, religious institutions and the private sector; (f) collaborations with academics on research projects, panel discussions and conferences; and (g) support from legal experts in drafting shadow reports and bringing forth test cases to challenge discriminatory anti-buggery laws.

LGBTQ groups have been instrumental in supporting the decriminalisation of buggery, which will be discussed further in the next section. While paying tribute to the late Caribbean gay rights activist Robert Carr, Kenita Placide, former co-founder and Executive Director of United and Strong and founder and current Executive Director of the Eastern Caribbean Alliance for Diversity and Equality (ECADE), clearly articulates the impact that LGBTQ activism has had on queer mobilisation in the region:

> We have challenged social norms and pushed through media advocacy, policy advocacy, documentation, capacity building, training and sensitisation nationally. We have also extended our work to regional partners. Among our flagship events are local and regional human rights sensitisation projects with police, sensitisation on LGBTQI issues and terminology with health workers and the Caribbean Women and Sexuality Diversity Conference (CWSDC). (Defenders, n.d.)

The vibrancy of Caribbean LGBTQ activism debunks the myth of the region being an impenetrable closet. Compared to previous generations, LGBTQ youth are now crucial change agents in the politicisation of gay and transgender rights in the region. For instance, in 2014, an alliance of LGBTQ youth organisations from Trinidad and Tobago, Barbados, Jamaica, St Lucia, Guyana and Belize came together to strategise on tackling discrimination in their societies and to advocate for a human rights mandate addressing LGBTQ rights in the Caribbean (Generation Change, 2014).

Given the diversity of LGBTQ subjectivities in the Caribbean and their varying positionalities, queer epistemic resistance is a site of contestation among different groups and approaches in a post-colonial landscape. For Medina (2013), contestation is part of a dialogic process for social change that allows for diverse perspectives to be heard and coalesced in upending oppression towards equitable social relations. He states: "What we need to maintain the ongoing possibility of resistance is epistemic friction. Epistemic friction involves the mutual contestation of differently normatively structured knowledges; it interrogates epistemic exclusions, disqualification, and hegemonies" (Medina, 2013, p. 281).

Epistemic friction has been apparent in Caribbean queer activism through the interplay of different persons, political persuasions and agendas. Attai (2017, p. 111) states that

> although these, and other, groups have invested extensively in public engagements on issues of wider acceptance of queer people, larger groups like

JFLAG, CAISO and SASOD have been criticised for their Afro-Caribbean, male-run and androcentric interventions, and an inattention to issues affecting lesbian women, trans men and trans masculine people in the region.

The author also cautions about LGBTQ perspectives that use nationalistic narratives or reproduce gay tropes to promote their cause. Epistemic friction also takes place in other areas of LGBTQ organising due to breakdowns in trust and communication with members and strategic alliances, competition with other civil society organisations for funding, financial and organisational instability, and state roadblocks.

Hermeneutical Friction, Homophobia and Political Backlash

The hermeneutical insensitivities against LGBTQ persons are not accidental but intentional, with leaders manipulating language and law to maintain a heteronormative status quo. Caribbean leaders perpetuate hermeneutical injustice on a structural level when the state does not respect the dignity and rights of *all* sexual citizens. This standpoint contradicts the image that they project to the international community as supporters of human rights. On multiple occasions, LGBTQ activists have pressured politicians to address the lack of equal protection and rights before the law for gay men, lesbians and transgender individuals. The responses have either been half-hearted or reflected an unwillingness to concretise efforts to challenge discriminatory laws and practices impacting the lives of sexual minorities.

Thus, a post-colonial heteropatriarchal logic prevails when it comes to sexual citizenship. For example, during proposed amendments to the Sexual Offences Bill in Jamaica in March 2009, then Prime Minister Bruce Golding was asked about violence against LGBTQ people and the impact of anti-buggery laws in influencing such harmful behaviour. Mr. Golding's hermeneutical arrogance was on full display when he stated that the government would not condone violence against LGBTQ people because of their "sexual preferences and lifestyle" while, in the same breath, reinforcing prejudice against gay people by upholding moral authority on the issue, contending, "We will never start peeping in anybody's bedroom to see what they are doing in privacy... but what we are not going to do is give official or legislative endorsement that holds that up and says this is a perfectly acceptable way to live" (Luton, 2009, p. 1). Mr. Golding's prejudices against LGBTQ people also extended to his unwillingness to have them serve in his cabinet. Based on discrimination in the law and the prejudicial and duplicitous viewpoint of some Caribbean leaders, same-sex intimacy is interpreted solely as a private matter, rendering it non-intelligible for collective understanding and unacceptable for public recognition.

In another example of political duplicity, in 2013, the president of B-GLAD wrote an open letter to Prime Minister Freundel Stuart of Barbados about the unfair treatment and violence that gay men and lesbians experience in society. The prime minister responded by pledging his commitment to ensuring that Barbados remains committed to protecting the human rights of all its citizens, including LGBTQ persons. While he demonstrated some sensitivity in acknowledging that LGBTQ persons should not be discriminated against in Barbadian society, he took no progressive action in his government to remedy the situation.

Mr. Stuart lacked hermeneutical virtue because he was not genuine in his intentions to challenge heteronormativity and homophobia in Barbadian society. Ro-Ann Mohammed (2017), a co-founder of B-GLAD, reports that this was made evident when the prime minister was asked about the anti-buggery law in Barbados while giving a lecture in Canada in 2016. To avoid a negative response from a foreign audience, Mr. Stuart minimised the severity of the laws on LGBTQ persons. To counter homonationalist condemnation from a Canadian populace, Mr. Stuart engaged in hermeneutical deception by stating that the dubious law deals with the crime of non-consensual sex in same-sex relationships, comparable to rape laws that deal with the crime of non-consensual sex in heterosexual relationships. He opined, "Therefore if the sex is consensual, there can't be a case. There is a lobby that is trying to get the government, trying to get successive governments in Barbados to decriminalize as they say homosexuality." "But you can only decriminalize something that is already a criminal offence" (Mann, 2017, para. 8).

Mr. Stuart equivocates on his statements and misrepresents laws to outsiders to make his country appear tolerant towards LGBTQ people by cleverly separating sexual identity from sexual conduct. He posits that being LGBTQ is not a crime in terms of identity, but engaging in "unnatural sexual acts" associated with homosexuality is punishable. At this point in time, buggery was a criminal offence in Barbados under the Sexual Offences Act 154(9) (2002), which reads: "Any person who commits buggery is guilty of an offence and is liable on conviction on indictment to imprisonment for life." This includes consensual anal sex between a man and a man or a man and a woman, even if done in private (Bulkan & Robinson, 2017). However, the law's vagueness produces hermeneutical deficits that state officials can purposely misinterpret to exercise discretion or to obfuscate the real intention of the law.

Mr. Stuart's mischaracterisation of the substance of the law did not hide his beliefs about the "unnaturalness" of homosexuality. He surmised:

> Those people who feel that we should create an environment where they can practice their lifestyles in public on high noon on a sunny day, or whatever, want even the very limited controls we have, removed. We have not reached a stage yet where we think that we want to do that. But we allow people to conduct their lives in accordance with their orientation or practices. (Mann, 2017, para. 9)

In reinforcing a post-colonial heteronormative script in shaping Barbadian modernity, Mr. Stuart invoked Christian values, and thus Christian citizenship, to justify a moralistic perspective (Lazarus, 2015) in hating the sin but loving the sinner, to set the parameters of which individuals are fully accepted as sexual citizens in his country. He opined: "They are values [Christian] that have helped to make Barbados the strong country that it has been over its entire history, but certainly over the last 50 years as an independent nation." Ironically, the Christian values that Mr. Stuart celebrates are the same values that the colonisers used to subjugate black bodies; post-colonial masters are now using them to suppress queer black bodies.

In examining the precarity of anti-buggery laws in the Bahamas, Jamaica and Trinidad and Tobago, Gaskins (2013) highlights the hermeneutical duplicity in the interpretation of these laws. He states that "despite the gender-neutrality of the law, sodomy laws are almost always mischaracterized as applying exclusively to homosexuals and it is usually animus toward homosexuals that prevents their repeal" (Gaskins, 2013, p. 436).

Mr. Stuart's attempt to downplay the severity of anti-buggery prohibitions (s. 9) is further complicated by Section 12.3 of the Sexual Offences Act (2002), which criminalises sexual indecency: "An act of serious indecency is an act, whether natural or unnatural by a person involving the use of the genital organs for the purpose of arousing or gratifying sexual desire." This section also threatens the sexual autonomy of all individuals, including gay, lesbian, bisexual and transgender persons, who engage in non-coital sexual activity.

In addition, Mr. Stuart's characterisation of the rape law is misleading and inadvertently exposes the discursiveness of the anti-buggery law if other laws in place criminalise non-consensual anal sex. The Sexual Offences Act 154 (s. 3) provides a gender-neutral definition of rape and does not limit its application to heterosexual scenarios, as Mr. Stuart suggests. Section 3.1 states:

> Any person who has sexual intercourse with another person without the consent of the other person and who knows that the other person does not consent to the intercourse [...] is liable on conviction on indictment to imprisonment for life.

Section 3.6 extends the scope of rape to include forceful entry with a penis or object of the vagina, anus or mouth. This inclusive definition means that individuals of any sexual orientation can be victims of rape by another person known or not known to them.

Moreover, as a lawyer and politician, Mr. Stuart engaged in wilful ignorance or "deception to obfuscate the truth" (Crawford, 2019, p. 123), and he did so by "actively ignoring the oppression of others and one's role in that exploitation" (Tuana, 2006, p. 11). Fortunately, the current prime minister of Barbados, the Honourable Mia Mottley, has long advocated overturning antiquated anti-buggery laws. She pushed for decriminalisation as Attorney General in 2003, but her effort was unsuccessful. However, the Barbados High Court struck down the anti-buggery law in 2022. Also during that year when Barbados became a Republic, the issue of legalising same-sex unions was discussed. The prime minister reiterated her position on sexual rights for all persons: "Nor can a society as tolerant as ours, allow itself to be 'blacklisted' for human and civil rights abuses or discrimination on the matter of how we treat human sexuality and relations" (Reid-Smith, 2020, p. 5).

LGBTQ Rights, Strategic Litigation
and Decolonising Caribbean Constitutionalism

Caribbean public spaces are contested political arenas for social movements, as they should be, because colonial and post-colonial realities collide, producing hermeneutical friction and confusion between LBGTQ groups, the state and other civil society groups, such as the church. These entities have different interests in rights as they relate to sexual citizenship, determining who belongs or does not belong to the nation. Intersectional politics between LGBTQ and feminist groups, as well as strategic litigation, have been instrumental in the epistemic resistance of marginalised populations, advancing the democratisation process in their societies.

As critical legal scholars, Bulkan & Robinson (2017, p. 231) understand the limitations of the law but see opportunities for legal redress for marginalised groups when there is a coalition of forces fighting for justice through *strategic litigation*. They note that "strategic litigation achieves the most when careful thought is given to which issues to litigate and when it is used alongside multiple political strategies and undertaken collaboratively" (2017, p. 231). In connecting critical legal theory and social advocacy to tackle discrimination against LGBTQ persons, three university public law lecturers, Tracy Robinson, Arif Balkan and Douglas Mendes, established the Faculty of Law, UWI Rights Advocacy Project (U-RAP), in 2009. The counter-public created by U-RAP through legal education, human rights and social justice was pivotal in two successful strategic litigation cases dealing with decriminalising buggery and vagrancy laws in Belize and Guyana, respectively.

Legal battles have not been easy for LGBTQ activists; Bulkan and Robinson (2017), p. 232) note, "Constitutional courts in the Caribbean have granted very limited access to LGBT organizations to bring claims on behalf of the communities they serve." Thus, LGBTQ organisations, such as J-FLAG, the Society Against Sexual Orientation Discrimination (SASOD), the United Belize Advocacy Movement (UNIBAM) and ECADE,[3] alongside other NGOs and U-RAP, must be credited for staying the course in mobilising and supporting plaintiffs to challenge discriminatory laws that criminalise same-sex intimacy and transgender expression. The strategic litigation, in testing the constitutionality of anti-buggery laws across the region, has been informed by local and international LGBTQ activism and the recognition of comparative human rights decisions in defending individual rights within the broader scope of universal rights enshrined in the fundamental rights and freedoms for citizens related to the right to privacy, life, liberty, security and protection of the law (Bulkan & Robinson, 2017).

Legislative reforms in other parliamentary democracies in the global North have resulted in precedent-setting cases on gender and LGBTQ

[3] Jamaica Forum for Lesbians, All-Sexuals and Gays (J-FLAG), Society Against Sexual Orientation Discrimination (SASOD), United Belize Advocacy Movement (UNIBAM), Eastern Caribbean Alliance for Diversity and Equality (ECADE).

rights that have influenced jurisprudence in the Commonwealth Caribbean (Elliott-Williams, 2019). In many ways, charter rights rely on social justice hermeneutics whereby equality based on sameness in identity is not the quintessential metric used in adjudicating discriminatory laws and practices. Instead, equity or the equivalence of rights, as suggested by Cornell (1992), rather than sameness or "likeness," is more critical in upholding the constitutional rights of persons. She argues for equivalence of rights, which "does not demand that the basis of equality be likeness to men" or any other established group (Cornell, 1992, p. 283). Moving away from an equality of likeness, Cornell (1992, p. 293) proposes an "equality of capability and well-being for women, homosexuals and others who live outside the dominant heterosexual matrix." In discussing Canadian jurisprudence, Ginn and Kindred (2017, p. 1) note that Charter cases do not require "majoritarian support" in adjudicating discrimination between parties. With this approach, minority rights are protected from the potential tyranny of the majority because the focus is on the breach of constitutional rights in disputes instead of the supremacy of one viewpoint.

In cases where there is a constitutional challenge to the legislation, Ginn and Kindred (2017, p. 19) note that it must be determined if the "state (represented by the legislature, an administrative decision maker or a court) balanced the constitutionally protected rights in a reasonable and proportionate way." The Caribbean Court of Justice (CCJ) and the High (Supreme) Courts of several Caribbean Commonwealth countries are decisively testing the constitutionality of colonial anti-buggery laws through adjudicative checks and balances. This is a progressive act in democratising sexual citizenship in the Caribbean to account for the sexual rights of LGBTQ persons.

Drawing on the landmark decision by the CCJ[4] in the *McEwan and Others v. Attorney General of Guyana* case in 2018, which ruled Guyana's crossing-dressing ban unconstitutional, Elliott-Williams (2019) argues that the CCJ, as the highest appellate court in the region, is playing a pivotal role in decolonising Caribbean constitutionalism. The CCJ carefully weighs the legitimacy and fairness of pre-existing colonial laws, such as vagrancy and anti-buggery laws, that contravene the Bill of Rights instead of seeing

---

[4] So far, only Belize, Barbados, Dominica and Guyana are under CCJ appellate jurisdiction. The other Caribbean Community members have retained the Judicial Committee of the Privy Council as their supreme appellate court.

such laws as absolute and untouchable based on the savings clause.[5] However, the CCJ's bolstered adjudicative role in overturning discriminatory laws cannot be seen in isolation from the decades of LGBTQ organising that helped politicise the issue and through the willingness of litigants to come forward to challenge unfair legislation.

Through epistemic resistance, Caribbean LGBTQ groups gathered, shared and disseminated collective knowledge and resources in creating counter-narratives and counter-publics to fight for their sexual rights. Their efforts simultaneously propelled actions to queer and decolonise sexual citizenship in the Commonwealth Caribbean. Thus, structural hermeneutical inequality born of homophobia and heterosexism (or any form of oppression) is not easily overcome by moral persuasion alone. Instead, it requires collective action on all levels in the fight for justice.

SAME-SEX INTIMACY AND THE ABOLISHMENT OF ANTI-BUGGERY AND VAGRANCY LAWS

Caleb Orozco v. The Attorney General of Belize, Supreme Court of Belize (2016)

It is important to note that while overt violence and discrimination against gay men and lesbians may be frowned upon by elites who present themselves as upstanding citizens of stable Caribbean democracies to the West, structural hermeneutical inequality in the form of homophobia and transphobia is operationalised in the law to maintain the status quo.

The 2016 decision by the Supreme Court of Belize in *Caleb Orozco v. The Attorney General of Belize* was precedent-setting in decriminalising buggery. Chief Justice Kenneth Benjamin delivered the decision in favour of the claimant, Mr. Caleb Orozco, a gay man. Although the decision was appealed in 2018, the Court of Appeal upheld the judgement on 30 December 2019. The judgement established seminal arguments in challenging discursive ideas and language in law used to police LGBTQ

---

[5] In discussing the purpose of the savings clause as a temporary measure to ensure a smooth transition from colonialism to independence, Kayne (2022, p. 1) notes that "a savings clause is a provision in a constitution which protects any law that was validly in force before the country's adoption of the constitution. It protects laws that might otherwise be struck down as unconstitutional on human rights grounds."

persons and criminalise non-procreative sex acts between consenting adults in private while appearing not to do so outrightly. It was also progressive in challenging religious and moral reasoning in law and addressing sexual rights based on both sexual conduct and sexual identity for sexual minorities. Moreover, the outcome of the Supreme Court and Court of Appeal decisions had a positive impact on other decriminalisation cases in other Caribbean Commonwealth countries.

Testimonial Justice and Overcoming Silences

The efficacy of strategic litigation cases relies on claimants coming forth to legally challenge the state for violating their rights. A critical part of this is testimonial justice in hearing and validating the voice of the oppressed in communicating their grievance (Fricker, 2007; Medina, 2013). The Supreme Court of Belize (SCB) hearing in 2016 was based on a complaint brought forth by Mr. Caleb Orozco and Human Dignity Trust as an interested party in 2010. In this constitutional challenge, the piece of legislation in contention was Section 53 of the Belize Criminal Code, which reads: "Every person who has carnal intercourse against the order of nature with any other person or animal shall be liable for imprisonment for ten years" (SCB, 2016).

As the Executive President of UNIBAM, Mr. Orozco and his lawyers argued that Section 53 of the Belize Criminal Code breaches the constitution because it targets and "criminalizes anal sex [as an unnatural act] between two consenting male adults in private" (SCB, 2016, p. 3). While on the surface, "every person who has carnal intercourse" does not seem to apply specifically to gay men or same-sex sexual activity in *intent*, because *all* persons are included in the prohibition against "unnatural sexual acts" (i.e. anal sex or non-procreative sexual acts), it is deceptive in its neutrality since gay men were being disproportionately surveilled and harassed by authorities because of their sexual orientation.

The claimant's testimony had a humanising effect, making his queer reality more intelligible to others. As an openly gay and gender non-conforming man, Mr. Orozco shared his experiences of being maligned for who he is and the struggles he faced in his personal and professional life as a result. Mr. Orozco articulated a grievance through a counter-narrative, exposing how Section 53 affected him and why it should be overturned.

Growing up, he was harassed and bullied for being "different" from other boys because he did not display stereotypical hegemonic masculine qualities. He noted, "In the late teenage years and early adulthood, many others sought to discourage and rid me of my effeminacy and presumed homosexuality to make me into a man" (SCB, 2016, p. 24). The ostracism he encountered from his peers during his teenage years contributed to feelings of loneliness and isolation. Like other gay youth living in a homophobic environment, Mr. Orozco hid his authentic self to protect himself from abuse and violence.

As an adult, he became more open and expressive about his gender and sexual identity, but this came at a price, as he constantly had to ward off harassment and threats to his life. In his role as the executive president of UNIBAM at that time, he provided HIV/AIDS and sexually transmitted infection services to those in need and served marginalised MSM and LGBTQ persons. This type of advocacy work meant that he was highly visible in various communities and other public spaces, heightening his vulnerability as a target of homophobia and transphobia. In addition, the testimony of a psychiatrist as an expert witness provided vital scientific evidence that homosexuality is not an illness requiring treatment (conversion theory) and that the abuse and violence perpetrated against MSM and gay men harm their mental health and quality of life.

Moreover, this case was an act of epistemic resistance on the part of Mr. Orozco and his supporters, challenging structural hermeneutical inequality in the law that reinforced heterosexism and homophobia. They pooled resources and utilised both experiential and scientific knowledge to oppose laws that wrongfully infringed on the rights of sexual minorities based on their sexual identity and conduct.

Hermeneutical Friction: The Supremacy of Law, Decoloniality and the Limits of Religious Persuasion

While the triangulated hermeneutical friction between the state, LGBTQ activists and Christian conservatism has been divisive, it has nonetheless been purposeful in debating sexual rights in the public domain rather than behind closed doors. Ginn and Kindred (2017) discuss the conflict between religious freedoms and LGBTQ equality rights in Canadian Charter cases, particularly regarding inclusion and discrimination in private and public institutions. They note that resolving discrimination cases in a pluralistic society is often complex due to the competing values and interests of different groups, so "balancing competing fundamental rights and freedoms must be done contextually, with a weighing of harms and benefits on each side" (Ginn & Kindred, 2017, p. 1). This challenge of mediating secular and religious viewpoints regarding homosexuality is also evident in Caribbean litigation due to implicit and explicit biases against LGBTQ persons manifested in culture, law and state.

Hermeneutical friction was evident in this case due to the balancing act of not breaching individuals' privacy rights on the one hand and protecting the public good on the other. The power and influence of Christian citizenship in nation-building within the Commonwealth Caribbean cannot be overstated. Regarding sexual citizenship, the church has played a significant role in transmitting heteropatriarchal beliefs not only in religious spaces but also through its engagement with the state and civil society. Lazarus (2020) notes that Christian groups are motivated to protect their special interests and the nation against what they perceive as the immorality of LGBTQ people. Lazarus (2020, p. 381) argues that "their interpretations of Judeo-Christianity are being presented as an objective, infallible truth that can provide answers and safeguard the interest of all of society." Religious intervention in dictating Christian sexual citizenship for everyone creates hermeneutical faults that obscure the reality of the rule of law, however imperfect, in ensuring the constitutional rights of citizens. With the fundamental rights and freedoms enshrined in the Constitution, Section 53 of the Criminal Code was on shaky ground despite the moral posturing of the religious leaders who were interested parties in this case against the decriminalisation of buggery.

The Preamble of the Belizean Constitution 1981 (rev. 2011) incorporates religious and secular principles that coalesce, creating a hermeneutical quandary for readers. It states:

> WHEREAS the people of Belize affirm that the Nation of Belize shall be founded upon principles which acknowledge the supremacy of God, faith in human rights and fundamental freedoms, the position of the family in a society of free men and free institutions, the dignity of the human person and the equal and inalienable rights with which all members of the human family are endowed by their Creator.

While the beginning of the preamble touts Belize as a nation upholding "the supremacy of God," deference to an omnipotent power is intertwined with respect for human rights, fundamental freedoms and the inalienable rights of individuals, which are believed to be "endowed by their Creator." The potential for misinterpretation of the preamble is apparent through the duplicitous language that obfuscates the separation of church and state and relegates the giver of "rights" to a celestial realm instead of to the domain of law, where religious freedoms also lie.

In the decision, Chief Justice Benjamin clarified the domain and limits of religion as it relates to the law and the protection of the common good. Instead of engaging in a theological debate about morality or trying to decipher the intent of the drafters of the preamble, the Chief Justice reaffirmed the separation of church and state and focused on whether there was a breach of constitutional rights. In clearing up interpretive confusion surrounding the language of the preamble, while the Chief Justice acknowledged that Belize is a Christian nation, he boldly questioned the placement of religious sentiments in the preamble since religious freedoms are guaranteed in the Constitution. He emphasised the state's secularism by noting that "the reference to God and the Creator does not import religious principles into the interpretation of the Constitution. The plain language of the Constitution must be given a liberal and purposive interpretation" (SCB, 2016, p. 22).

This cogent explanation of the separation between church and state left no room for hermeneutical confusion about the purpose of the preamble as an overture to the Constitution instead of being the body of law itself. These counter-arguments against ecclesiastical hermeneutics are necessary for destabilising religious heteropatriarchal beliefs (both colonial and neocolonial) that are embedded in law to police and delegitimise black queer bodies (Lazarus, 2015).

Since this case was about a constitutional challenge to legislation and not about competing charter rights between religious freedoms and gay rights (such as churches being forced to accept gay clergy or congregants against their religious beliefs), Section 53 was not tenable. While the preamble vacillated between secular and religious sentiments, Part II of the Protection of the Fundamental Rights and Freedoms (Sections 3 to 19) of the Constitution (rev. 2011) fully grants an individual fundamental rights and freedoms regardless of race, place of origin, political opinion, colour, creed or sex. Section 3 lists the core rights embedded in liberal democracies:

(a) life, liberty, security of the person, and the protection of the law;
(b) freedom of conscience, of expression and of assembly and association;
(c) protection for his family life, his personal privacy, the privacy of his home and other property and recognition of his human dignity; and
(d) protection from arbitrary deprivation of property.

Mr. Orozco brought forth the claim that Section 53 violated his fundamental rights under the Constitution Sections 3(c), 6(1), 11, 12 and 14(1), which cover the fundamental rights and freedoms, protection of the law, freedom of conscience, freedom of expression and the protection of the right to privacy (protection from arbitrary and unlawful interference of privacy in the home or private life), respectively. In the following sections, I will analyse the seminal arguments of the Supreme Court of Belize that provide a more progressive legal hermeneutical position influenced by human rights law in expanding sexual rights for LGBTQ people.

Sexuality Identity, Human Dignity and the Right to Privacy

Both the interpretation and the impact of laws matter when challenging structural hermeneutical injustice, which was apparent in the case of *Caleb Orozco v. The Attorney General of Belize*. The arguments in favour of decriminalisation were bold and comprehensive, supporting the personhood and fundamental rights and freedoms of LGBTQ people. The Chief Justice stated that the Constitution of Belize is a "living instrument," not fixed in space and time but evolving with the needs and values of society. This dynamic nature of the Constitution aligns with shared values of cooperation, democracy and human rights within an international framework.

Belize, as a signatory of several international covenants, such as the Charter of the Organization of American States, the Inter-American Democratic Charter, the United Nations Charter and the Universal Declaration of Human Rights,[6] aligns its domestic laws with its international obligations. It states that "the Belize Constitution owes its provenance to the European Convention on Human Rights which in turn is influenced by the UN Declaration of Human Rights" (SCB, 2016, p. 24).

---

[6]International   Conventions,   https://mfa.gov.bz/multilateral-international-conventions/—Ministry of Foreign Affairs, Foreign Trade and Immigration. Government of Belize.

For post-independent Caribbean countries, friction exists between colonial vestiges and post-colonial realities and partnerships within a globalised world as they forge new principles and laws distinctive to their citizenry and socio-political institutions. In overcoming hermeneutical injustice in the law, words matter, and the language must evolve in a semiotic order (Medina, 2013) to promote diversity, equity and inclusion. The Supreme Court of Belize recognised this and concurred that the term "sex" includes sexual orientation, drawing on the UN Human Rights Committee definition used in the *Toonen v. Australia Communications* (1992) case that overturned anti-sodomy laws in Australia.

Adopting a sociological perspective in referring to "sex"—incorporating the biological, physical and intimate aspects of personal embodiment beyond the sex/gender binary—established the central basis for addressing the violation of the claimant's rights to human dignity and privacy based on his *sexual identity*.

Mr. Orozco argued that Section 53 of the criminal code infringed on his *human dignity* (Section 3(a)) by demeaning him as a gay man by "categorising consensual male homosexual acts in private with forced intercourse, sex with minors and sex with animals" (SCB, 2016, p. 26). This denial of human dignity constitutes a form of epistemic injustice, as Fricker (2007) explains, because when a knower is discriminated against, their oppressive situation depletes their self-worth or worthiness. The emotional and psychological stress they experience obscures their ability to recognise their value to themselves and others. Consequently, homophobia harms the personal dignity of sexual minorities, who are judged as less than heterosexuals, by undermining their self-worth, self-respect and the appreciation that they seek from others (Baker & Lucas, 2017).

The violation of human dignity is not always apparent because discriminatory acts can be both explicit (overt) and implicit (covert). In this case, the prohibition against anal sex in Section 53 applied to "every person who has carnal intercourse" and not specifically to LGBTQ people. However, the High Court saw through the veiled neutrality of the law concerning human dignity. Drawing on Mr. Orozco's testimony and a gay rights case in South Africa, Chief Justice Benjamin reasoned that the claimant's right to human dignity (Section 3(c)) was breached by Section 53. Although the law appeared neutral, "the impact on the dignity of a homosexual man is disproportionate given the deep stigmatisation caused by them being the primary targets" (SCB, 2016, p. 24).

In this way, legal and social justice hermeneutics converged, considering equity (fairness) rather than equality (sameness) to determine the impact of the anti-buggery law. This law disadvantaged LGBTQ people more than heterosexuals concerning human dignity based on differences in sexual identity and orientation. Ultimately, the full personhood of sexual minorities could no longer be denied.

Regarding *privacy and sexual conduct*, the claimant's arguments were clear and crucial to the broader discussion of sexual rights that limit state interference in people's private lives and bedrooms. Mr. Orozco contended that Section 53 explicitly violates a person's right to privacy under Section 14(1):

> A person should not be subjected to arbitrary and unlawful interference with his privacy, family, home, or correspondence, nor to unlawful attacks on his honour and reputation. The private and family life, home and the personal correspondence of every person shall be respected. (SCB, 2016)

Unsurprisingly, hermeneutical friction resurfaced between religious interpretations of morality related to homosexuality, individual rights and protecting public interests. The defendant argued that there are limits to the right to privacy if public safety, health and morality are threatened. Church leaders were present as interested parties opposing the decriminalisation effort. Accordingly, Williams et al. (2020, p. 730) argue that "religious ideologies and strong ties between political parties and Christian institutions in the Caribbean are the most significant obstacles to advancing decriminalization or human rights for LGBT people." Armed with heterosexist colonial Christian beliefs, the church leaders argued that Belize is a Christian nation and that homosexuality goes against God's will. Furthermore, they feared that decriminalisation would lead to more vice or sinful behaviour in society. Lazarus (2015) critically notes that Caribbean religious civil society organisations view themselves as gatekeepers of morality, safeguarding their nations against disorder and debauchery and promoting heteronormative sexual citizenship. Williams et al. (2020, p. 730) add that

> the current homophobic rhetoric delivered from church pulpits continues the work of imposing sexual control over the population and, in contemporary times, that influence is augmented by American religious groups stoking anti-homophobia rhetoric as part of their spread into the Caribbean region.

This was evident in this case, where religious supporters of Section 53 argued that morality is the dominion of "God," which the law and state should abide by to protect the "common good."

Whether due to local or foreign conservative influences, the state and elected officials have been complicit in supporting fundamentalist religious viewpoints against sexual minorities. Some politicians have masterfully engaged in wilful ignorance by weaponising anti-buggery laws to intimidate LGBTQ people (Crawford, 2019) while knowing they cannot forcibly interfere with sexual conduct between consenting adults in private based on privacy rights enshrined in the Constitution under Section 14(1). Chief Justice Benjamin challenged the majoritarian religious interpretation of homosexuality as a danger to the public good by stating, "No evidence has been presented as to the real likelihood of such harm" (SCB, 2016, p. 31). Ultimately, belief and moral persuasion alone by clerics were insufficient to deem homosexuality a threat to society. Section 53 breached the claimant's right to privacy and failed another test for it to be upheld.

Freedom of Sexual Expression and Equality Under the Law

The claimant argued that Section 53 violated his rights to human dignity and privacy under the Constitution, seeking a more inclusive understanding of personhood that protected the rights of LGBTQ people. Mr. Orozco asserted that Section 53 violated his freedom of expression guaranteed under Section 3(b) and Section 12(1) of the Constitution because it contravened the "diversity and difference of opinion contemplated in the Constitution" (SCB, 2016, p. 34). Since freedom of expression was conceptualised broadly to include different mediums of receiving and communicating ideas, opinions and information without interference from the state and others unless it threatened public safety, Chief Justice Benjamin drew on precedent cases from the United Kingdom and Canada to surmise that "freedom of expression is one of the pillars of a democratic society" (SCB, 2016, p. 34).

Freedom of expression is a critical component of epistemic justice because, through testimonial exchanges, "communicators participate as knowers" of their own lives, which can be shared with others. Medina notes that testimonial exchanges can be communicated through written, oral or creative modalities, both directly and indirectly. In understanding the importance of freedom of expression to Mr. Orozco's personhood, the High Court considered sexuality (identity, expression and conduct) as a form of personal expression because "it attempts to convey meaning" (SCB, 2016, p. 34). Since the claimant did not include freedom of expression in his argument, the High Court took a hermeneutically virtuous position to make this point. This helps further mitigate sexual unfairness by recognising how freedom of expression is an intrinsic part of a democracy, allowing individuals to be themselves without fear of reprisals.

The final blow in striking down Section 53 came when it was determined to breach the claimant's right to equality under and before the law. Since earlier arguments established that Section 53 violated Mr. Orozco's core fundamental rights and freedoms based on sex as set out in Section 3 of the Constitution, protection from discriminatory treatment by law and state as outlined in Sections 6(1) and 16 was also considered. Under Section 6(1) of the Constitution: "All persons are equal before the law and are entitled without discrimination to equal protection of the law" (SCB, 2016, p. 36). This was complemented by Section 16, which "confers protection against discriminatory laws and treatment by person or authority against another person" (SCB, 2016, p. 35).

Overcoming structural hermeneutical inequality requires fairness in laws, adjudication and due process. Chief Justice Benjamin reiterated that Section 53 did not meet this standard set out in Sections 6(1) and 16 of the Constitution, rendering it unconstitutional: "As previously iterated, inasmuch as Section 53 is framed in gender-neutral language, the evidence demonstrates that it is discriminatory in its effect. The claimant has shown that he has been rendered a criminal by virtue of his homosexuality" (SCB, 2016, p. 35). The epistemic advantages garnered from Mr. Orozco's testimony and precedent-setting rulings in international law, which recognise sexual orientation as a form of sex discrimination, secured a successful outcome for sexual minorities in abolishing anti-buggery laws in Belize.

Implications of Orozco Case in the Re-Interpretation of Sexual Citizenship

Caleb Orozco v. The Attorney General of Belize was a landmark decision in Caribbean jurisprudence that brought to the forefront the importance of strategic litigation and social justice hermeneutics in challenging discriminatory anti-buggery laws that infringe on the sexual rights of LGBTQ persons. The epistemic resistance of LGBTQ activists set this case in motion, and its successful outcome ignited other decriminalisation cases. This precedent-setting case generated new hermeneutical collective understandings and advantages in valuing the personhood of LGBTQ people and protecting their rights against discrimination in law and state. It contributed to a greater collective understanding of the issue that mobilised resources and people to fight battles in other Commonwealth Caribbean countries. Through sound judicial reasoning informed by human rights laws and social justice hermeneutics, the decoloniality and queering of sexual citizenship were made possible.

Several key points emerged from this decision: (a) it upheld constitutionalism by emphasising the rule of law and the separation between church and state; (b) the court scrutinised discursive language in legislation that, while neutrally framed, had disproportionately adverse effects on LGBTQ people; (c) diversity, equity and inclusion were considered in the interpretation of fundamental rights and freedoms to cover different categories of people; and (d) sex was defined broadly to encompass physical attributes, sexual identity and sexual orientation, acknowledging the biological, social and cultural markers of sexuality to support a more comprehensive notion of sexual citizenship.

Decriminalisation and Expanding Rights for Sexual Minorities: Jason Jones v. The Attorney General of Trinidad and Tobago

The central arguments in the Orozco case proved beneficial in the case of *Jason Jones v. the Attorney General of Trinidad and Tobago.* On 12 April 2018, the High Court of Trinidad and Tobago found the anti-buggery law and other prohibitions against same-sex intimacy unconstitutional. The Jason Jones decision offered a more comprehensive critique of the unconstitutionality of anti-buggery laws and examined how LGBTQ people face discrimination in various aspects of intimate life, including in creating a family. This case fostered a greater collective understanding of LGBTQ rights, helping to mobilise resources and people to fight similar battles in other Commonwealth Caribbean countries. These efforts led to successful decriminalisation cases in Antigua and Barbuda (2022), St. Kitts and Nevis (2022), and Barbados (2022).

Unnatural Acts and the Homosexual Outsider

The anti-buggery law in Trinidad and Tobago had a broader scope in its categorisation of "unnatural acts" and imposed harsher penalties for violations, making structural hermeneutical inequalities against gay people particularly stringent and difficult to overcome. Alexander (1994) notes that the redrafting of morality in Trinidadian law during the post-independence period further entrenched heteropatriarchal sexual citizenship, marginalising lesbians and other sexual minorities. The anti-buggery law, found in Section 13 of the Sexual Offences Act, criminalised buggery, or sex per anum, between males and males and between males and females, with those convicted facing up to 25 years in prison. Additionally, Section 16 prohibited serious indecency, defined as "an act other than sexual intercourse (whether natural or unnatural), by a person involving the use of the genitalia organ for the purpose of arousing or gratifying sexual desire" (High Court of Justice [HCJ], Trinidad and Tobago [T&T], 2018, p. 4). This violation carried a penalty of five years in prison. However, the exemption for serious indecency applied only to private acts between consenting adults of the opposite sex and children, making LGBTQ people or those engaged in same-sex intimacy the prime targets for punishment for consensual, non-procreative sexual acts. This constituted blatant sex discrimination based on sexual orientation. Given the magnitude and importance of this case, the High Court of Trinidad and Tobago thoroughly deliberated on it, and Chief Justice Devindra Rampersad delivered a favourable outcome for the claimant.

Mr. Jones' role as the claimant in this decriminalisation case was driven by his personal and collective activism in HIV/AIDS prevention and the fight for LGBTQ rights, both in Trinidad and Tobago and London. As a Trinidadian-born openly gay man, Mr. Jones moved to London in his early 20 s in 1985 to seek better professional opportunities and escape the homophobic environment in the small island state. While living abroad, he occasionally visited Trinidad and Tobago, and in 1992, he had an extended stay that lasted a few years before he returned to London in 1996. During his time abroad, he was a member of an advocacy group that fought for the "rights of same-sex partners of UK citizens overseas to be granted residency" (Jade, 2020, p. 2).

With intense lobbying during an election year, Mr. Jones and his partner were among the 40 test cases that pressured the incoming political party to pass a law providing a pathway to residency and citizenship for individuals in foreign/UK same-sex unions. The law passed, and his advocacy in this case would be fortuitous 21 years later in his country of birth when he became the claimant to challenge the anti-buggery law in Trinidad in 2017.

In the decriminalisation case in 2017, Mr. Jones testified that he feared for his life and that his ability to live a safe and healthy private life with his partner was compromised due to a threatening environment. He also recounted that his gay friends were frustrated by the lack of seriousness and action demonstrated by the police when they reported theft and violence against them by a group of men. Accordingly, Mr. Jones argued that Sections 13 and 16 of the Sexual Offences Act violated his fundamental rights and freedoms under Section 4 of the Constitution: (a) the right to liberty and security of person and due process; (b) the right to equality under and before the law; (c) the right to respect for private and family life; and (d) the right to freedom of thought and expression.

He declared that the prohibitive law was unreasonable because it disproportionately sanctioned him as a gay man, discriminated against him based on his sexual orientation, and discriminated against him based on the sex/gender identity of his partner (HCJ, T&T, 2018, pp. 27–28). The first and second arguments were also introduced in the Belizean decriminalisation case, but the third, concerning gay relationships or partnerships, was not fully developed. By directly challenging a heteronormative nuclear family model that excludes LGBTQ people, a more nuanced hermeneutical understanding of intimacy was presented. This included affective and familial bonds rather than simply focusing on sexual conduct, thereby broadening the realm of sexual citizenship for same-sex couples.

Constitutionalism and Sexual Rights

The defendant's attempt to invoke the savings clause to exempt Section 13 from judicial review was denied, effectively challenging colonial precedent. Chief Justice Rampersad reasoned that Section 13 breached the constitutional rights of the claimant under Section 4 (protection of core fundamental rights and freedoms) and Section 5(2)b (prohibition against Parliament instituting cruel or unusual treatment as punishment). In its ruling, the High Court adopted a cosmopolitan approach by drawing on international treatises and precedent-setting cases, including the Belizean decriminalisation case, while also considering changing social attitudes about homosexuality. Notably, the Court upheld constitutional supremacy, displacing the preponderance of legislative authority established under colonialism:

There is no doubt that the Bill of Rights in Trinidad and Tobago were born out of the Canadian experience and tweaked for local assimilation. Canada has adopted this method of shifting the burden on to Parliament rather than on to the applicant in respect of justification. To cling to the presumption of constitutionality is, to my mind, and with the greatest respect, a symbol of a further clinging to the vestiges of the colonial idea of the supremacy of Parliament which has, to my mind, been supplanted by the Constitution. (HCJ, T&T, 2018, p. 32)

The Constitution reigns supreme over past (colonial) laws, and the burden of proof falls on the defendant (Attorney General) to prove the reasonableness of the contested law. This principle, along with Trinidad and Tobago's legal alignment with other Commonwealth countries, set the stage for removing Section 13.

Like the *Orozco v. Attorney General* case, Chief Justice Rampersad concluded that the claimant's human dignity was violated due to bigotry against him as a gay man. Mr. Jones did not enjoy the presumption of innocence under the law compared to heterosexual men, as he was outrightly perceived as a criminal. Along with the clear violation of due process, the discriminatory law provided elected leaders and the state with the authority to discriminate against LGBTQ people engaged in consensual anal sex. The Chief Justice surmised:

Unlike heterosexual citizens, the claimant is treated differently under the law by reason of his sexual orientation in respect of the manner in which he expresses his love and affection. By engaging in that expression consensually, he is liable to be imprisoned for a term of up to 25 years—that amounts to a term of life imprisonment virtually—and the same does not apply to a heterosexual male unless he engages in intercourse per anum. The Act impinges on the right of the individual to equality before the law and the protection of the law. (HCJ, T&T, 2018, p. 30)

This astute judicial and social justice interpretation exposes how the law reproduces social inequities between different groups or classes of people.

State and Church Divided

One of the strengths of the ruling is the thorough attention the High Court gave to the history of anti-buggery laws in Trinidad and Tobago, tracing their evolution from the colonial to the post-independence period.

The Court also clarified any confusion between the secular nature of the state and religious edicts on morality that impact constitutional rights. While opponents criticise decriminalisation as interference by Western liberals, they ironically overlook that these prohibitive laws were initially imposed by colonial powers to control working-class and queer black people. Thus, the legislation was not created with the benefit of the local populace in mind. In a multicultural society like Trinidad and Tobago, with its diverse religious traditions, Chief Justice Rampersad emphasised the supremacy of the law over religion in this case: "This is not a case about religious and moral beliefs but is one about the inalienable rights of a citizen under the Republican Constitution of Trinidad and Tobago; any citizen; all citizens" (HCJ, T&T, 2018, p. 58). This statement firmly establishes the separation of church and state and highlights the importance of protecting minority rights within a majoritarian framework, including the rights of LGBTQ people and other sexual minorities.

In reviewing the claimant's case, similar to the Orozco case, Chief Justice Rampersad exposed the structural deficiencies in the law rooted in colonial religious biases. He addressed the defendant directly:

> Therefore, when counsel for the defendant stated that this was not a matter about homosexuality, this court respectfully disagrees. The retention of the law seems to have everything to do with homosexuality and the colonial abhorrence to the practice which has been retained by the State in its separate identification and isolation in the very onerous provision under the Act. (HCJ, T&T, 2018, p. 49)

Referencing the Supreme Court of India's decriminalisation case (*Puttaswamy v. Union of India*), the origin and intent of the law were clarified. He noted that in thirteenth-century England, anti-sodomy laws emerged from biblical accounts of Sodom and Gomorrah, a place "associated with depravity, unnaturalness and substandard moral and spiritual values and existence" (HCJ, T&T, 2018, p. 49). England criminalised sodomy or sex other than "heterosexual penile-vaginal" intercourse, including oral sex, which was later revised to include only buggery (anal sex). The punishment under the Buggery Act of 1533 was death by hanging, and this law was re-acted in 1563 and transmitted to British colonies.

The prohibitions against non-procreative sexual activity under the Sexual Offences Act underwent various iterations during Trinidad's colonial and post-colonial history, with criminalisation and sentencing for buggery and other non-procreative sexual activities (gross indecency). The constant factor was that harsher penalties were designated for LGBTQ people. Thus, the historical account of buggery in the decision was purposeful in separating myth from fact and law from religion, exposing the heteropatriarchal religious prejudices in the law (Sections 13 and 16) that breach the claimant's constitutional rights. Chief Justice Rampersad concluded, "As this history illustrates, the offence born out of the Christian church's patriarchal moral jurisdiction and yielded, and continues to yield, serious consequences statutorily" (HCJ, T&T, 2018, p. 12). He further denounced the hermeneutical deficiencies of the law by questioning its relevancy as a safeguard against sodomy, noting that rape laws are broad enough to cover non-consensual sex per anum. This directly challenges politicians who justify upholding anti-buggery laws as a measure to prosecute acts of forced anal sex.

The Right to Privacy Includes Same-Sex Intimacy and Family Life

Establishing a family is a significant area of intimacy for LGBTQ individuals, where structural hermeneutical injustice—manifested through heterosexism and homophobia— obscures their affective bonds. The claimant argued that the right to privacy in engaging in consensual sexual acts inherently includes the right to have a family life. He expressed concerns that, as gay men, he and his partner could not freely and safely live in Trinidad and plan a family. This strategic move was effective in humanising same-sex relationships and advocating for comprehensive sexual rights by moving beyond harmful stereotypes of homosexual depravity and hedonism. Chief Justice Rampersad concurred with this view, stating:

> The claimant, and others who express their sexual orientation in a similar way, cannot lawfully live their life, their private life, nor can they choose their life partners or create the families that they wish. To do so would be to incur the possibility of being branded a criminal. The Act impinges on the right to respect for a private and family life. (HCJ, T&T, 2018, p. 30)

The infringement on the claimant's right to privacy, relating to personal intimacies and bodily autonomy, was evident. However, the defendant questioned the right to family life and failed to provide substantive evidence to justify maintaining the law, aside from citing a desire "to maintain traditional family and values that represent society" (HCJ, T&T, 2018, p. 47). This was supposedly mentioned to clarify the law and ensure it was upheld in its entirety, including Section 16, which addressed serious indecency. However, it was noted that Section 16 primarily targeted lesbians and was seldom enforced against consenting homosexual men. This observation aligns with Alexander's (1991) argument that the redrafting of morality in the Sexual Offences Act functioned is a heteropatriarchal mechanism to control women's bodies—both sexual and reproductive—by restricting their erotic autonomy.

The court addressed the hermeneutical flaws in the defendant's arguments regarding the claimant's right to family in the following manner. Firstly, Chief Justice Rampersad acknowledged the significance of the "traditional family" (albeit based on a Eurocentric worldview), though he did not see this unit as fixed or unchanging over time. He illustrated this point by referencing the prevalence of single-parent families in the Caribbean: "There is no doubt that maintaining traditional family values that represent society are important concepts, but those words have now to be adapted to a different world than medieval and Victorian times" (HCJ, T&T, 2018, p. 53). While this perspective represents a progressive shift away from a heteronormative nuclear family model and a structural-functionalist view of family, it faced opposition from those concerned that an inclusive understanding of family might lead to marriage equality. Nonetheless, recognising the right to family life for same-sex couples affirms queer family structures. It embraces a broader, more inclusive understanding of Caribbean family life that transcends a heteronormative nuclear family model rooted in Enlightenment gender ideologies. This topic will be further explored in Chap. 5, where I will discuss the decolonisation and queering of Caribbean family structures.

Secondly, the High Court criticised the defendant for attempting to uphold Section 16, which was discriminatory towards LGBTQ people, despite the argument that it did not infringe on the claimant's rights because it was rarely enforced against gay men. Chief Justice Rampersad condemned this approach as a "'big stick' over a minority to try to enforce a portion of society's morality over it" (HCJ, T&T, 2018, p. 47). The Court deemed the law unnecessary if it was unenforceable, considering it more "vindicative than protective or curative in any manner" (HCJ, T&T, 2018, p. 48). Ultimately, based on the evidence presented by the claimant and the precedent-setting decriminalisation cases, the High Court of Trinidad and Tobago ruled that "Sections 13 and 16 of the Act are unconstitutional, illegal, null, void, invalid and are of no effect to the extent that these laws criminalise any acts constituting consensual sexual conduct between adults" (HCJ, T&T, 2018, p. 54).

The Way Forward: Strategic Litigation and Hermeneutical Justice

The central arguments in the *Jones v. Attorney General of Trinidad and Tobago* and the *Caleb Orozco v. The Attorney General of Belize* Supreme Court decisions were informed by social justice hermeneutics in the nullification of anti-buggery laws. Sexual citizenship was reframed to recognise homosexual expression and human dignity, along with protected rights of privacy, due process and consensual sexual activity regardless of one's sexual orientation. This dual effect of decolonising and queering sexual citizenship involved recognising and protecting the civil rights of homosexual sexual citizens. The role of the church and legislators in using the law to enforce morality based on heteropatriarchal beliefs was destabilised by judicial hermeneutical justice.[7]

The strength of Caribbean post-colonial constitutionalism was demonstrated through the judiciary's use of domestic and international laws to uphold democratic principles and the rule of law. The courts viewed the claimants as credible speakers, voicing their grievances to challenge discrimination (heterosexism and homophobia) in law, state and society. This successful fight for minority rights protection was achieved through the epistemic resistance of claimants, lawyers, LGBTQ activists, and regional and international NGOs. These groups courageously used collective resources and knowledge to ensure marginal voices were heard and

---

[7] Not surprisingly, different faith leaders have publicly denounced the nullification of anti-buggery laws. For example, before and after the ruling of the Supreme Court of Belize on the *Caleb Orozco v. AG of Belize* case, which decriminalised buggery in Belize in 2018 after a q0-year battle for the claimants, the Christian evangelicals mobilised allies in civil society as well as foreign religious conservative supporters to pressure the government to uphold buggery laws to protect the citizenry from morally corrupting "unnatural behaviour," which they felt could also open up the floodgates for expanded gay rights such as gay marriage (Lazarus, 2020). Similarly, after the High Court of Trinidad and Tobago struck down the anti-buggery law for its unconstitutionality in 2020, faith leaders from different religious denominations banded together in their shared bigotry to protest the decision. One leader opined, "The fabric of society was at risk" (Williams-Sambrano, 2018, p. 1).

discriminatory laws and systems of oppression were challenged. The seminal arguments from these two decriminalisation cases produced a ripple effect in three other small island states in 2022. It is only a matter of time before we see which of the remaining six countries will be next.

Transgender Identities and Abolishing the Cross-Dressing Ban: McEwan and Others v. Attorney General of Guyana (2010; 2013; 2017; 2018)

While significant public debate has long surrounded the criminalisation or decriminalisation of buggery, the discussion about transgender identity and rights has more recently gained traction in popular culture. The civil rights of Caribbean transgender persons have been marred by the gender binary and heteronormativity upheld in law, state and wider society, leaving these individuals vulnerable to stigma, discrimination, and violence (Out Right Action International, 2021). The *McEwan v. Attorney General of Guyana* case was pivotal in bringing the struggles of this marginalised group to the forefront.

Prior to the CCJ's 2018 decision that found Guyana's cross-dressing ban unconstitutional, the law prohibited gender-variant dressing and expression. This ban had been upheld by the High Court and the Court of Appeal in Guyana before being overturned by the CCJ. Under the 1893 vagrancy law, Section 153(I)(XLVII) of the Summary Jurisdiction (Offences) Act, it was a criminal offence for a man to dress in women's attire or a woman to dress in men's attire in public for "any improper purpose." Elliott-Williams (2019, p. 744) elucidates the colonial intent of the anti-cross-dressing law and others like it in controlling black and brown bodies:

> The impugned measure was part of a suite of colonial era vagrancy laws, meant to wrest personal liberty from ex-saves and indentured labourers for reasons, which included the enforcement of Victorian values to ensure continued plantation production by controlling the use of public space.

The anti-cross-dressing law reflects structural hermeneutical inequality against transgender and gender non-conforming individuals, whose identities and expressions are rendered incomprehensible within a social system predicated on cisgender identities and heteronormativity. Caribbean transgender persons faced an arduous legal battle in seeking justice in this discriminatory gender identity and expression case. The lengthy legal process heightened visibility and coincided with efforts to decriminalise bugger in other Commonwealth Caribbean countries. This period was pivotal for LGBTQ organising in the region, galvanising people and resources through a surge of epistemic resistance. This resistance created queer counter-publics and utilised strategic litigation to seek gender and sexual justice.

Transgender Bodies Criminalised as Disorderly

The constitutional challenge against the cross-dressing ban in 2010 was spearheaded by four transgender individuals, Quincy McEwan, Seon Clarke, Joseph Fraser and Seyon Persaud, who were born male but identified with feminine attributes. On 6–7 February 2009, these four, along with three others, were socialising in the city centre of Georgetown while dressed in female attire. The visibility that accompanies being transgender made it difficult for them to conceal their identities, a survival strategy to avoid victimisation from the public and punishment from the state. Due to their genderqueer expression, they were surveilled and treated as a spectacle by police officers and some objecting onlookers who ridiculed and attacked them. Despite their attempts to defend themselves, "at the time of the arrests, McEwan was dressed in a pink shirt and a pair of tights and Clarke was wearing slippers and a skirt. A few hours later, Fraser and Persaud who were dressed in skirts and were wearing wigs, were also arrested by the police" (CCJ Press Release, 2018).[8]

While detained, they were denied an explanation of the charges against them, medical attention and the right to legal counsel. When their case was heard by Chief Magistrate (Ag.) Madame Melissa Robertson, several days later, the accused were charged with "loitering and wearing female attire" (High Court of Guyana, 2013, p. 13). This abuse of power and violation of due process by law enforcement was compounded by the hermeneutical faults in the law, which criminalised their genderqueer expression. The systemic discrimination these transgender individuals faced was challenging to overcome alone. Despite believing they had done nothing wrong, they pleaded guilty to the charges and were fined to avoid severe penalties.

Chief Magistrate (Ag.) Robertson showed no empathy towards the individuals, given their predicament. Instead, she demonstrated both hermeneutical ignorance and arrogance about transgender realities by telling

[8] https://today.caricom.org/2018/11/13/ccj-declares-guyana-cross-dressing-law-unconstitutional/

them that "they were confused about their sexuality, and they were men, not women, and advised them to go to church" (High Court of Guyana, 2013, p. 11). Transgender persons were dismissed as credible subjects, perceived as deviant and disorderly for contravening dominant categorisations of sex and gender associated with cisgender heterosexual individuals. Rather than receiving justice, they were offered religious advice to cure their supposed affliction, ironically from the judiciary, which is supposed to be a secular entity.

Constitutional Challenge to the Cross-Dressing Ban

Demonstrating courage and fortitude, in *McEwan and Others v. Attorney General of Guyana* (2010), the four previously charged transgender persons, along with the SASOD and assisted by their counsel of university lawyers, Dr. Arif Bulkan and Gino Persaud, filed a constitutional action against the government/state. This action listed nine items of contention, questioning the violation of the applicants' right to the protection of the law, the breach of fair arrest and detainment procedures due to mistreatment by law enforcement, and the charges lodged against them (including the request for damages). Additionally, it directly challenged the cross-dressing ban, Section 153(I)(XLVII) of the Summary Jurisdiction (Offences) Act, based on sex and gender discrimination and due to the obstruction of the rule of law because of its vague language and ubiquitous intent.

The two-pronged approach to this strategic litigation by the applicants was crucial in uncovering the different ways the constitutional rights of transgender persons were being violated due to discriminatory laws and judicial hermeneutical prejudices. Once again, there were inconsistencies with the maintenance of colonial laws alongside post-colonial constitutionalism. Bulkan and Robinson (2017) note that under the legal counsel of university lawyers from U-RAP and another lawyer, the applicants requested that personhood in the Constitution be expanded to recognise transgender individuals as deserving of their civil rights, which include freedom of expression and their right to protection under the law.

Like other Commonwealth Caribbean countries, Guyana's Constitution (2016) adheres to democratic principles and secularism, which are clearly defined in the preamble:

> Forge a system of governance that promotes concerted effort and broad-based participation in national decision-making in order to develop a viable economy and a harmonious community based on democratic values, social justice, fundamental human rights, and the rule of law.

In addition, the Protection of the Fundamental Rights and Freedoms of the Individual (introduced in Article 40 and laid out in Title I, Part 2 of the Constitution) is a suite of provisions that include, but are not limited to, the protection of life, the right to personal liberty (which includes proper arrest and detainment procedures), protection from inhumane treatment, protection of the law, and protection of freedom of conscience, expression and assembly. Section 149, which was heavily weighted in the case, dealt with "protection from discrimination on the grounds of race, place of origin, political opinions, colour, creed, age, disability, marital status, sex, gender, language, birth, social class, pregnancy, religion, conscience, belief or culture [...]," among others (Guyana's Constitution of 1980, 2016).

The case was heard in the High Court of Guyana in 2013, and Chief Justice Ian N. Chang delivered the decision. Chief Justice Chang swiftly deliberated on the applicants' claim that the police breached their rights and ruled that "the applicants' right to be informed of the reason for their arrest and detention was infringed by the State through the agency of the Police" (p. 32). Each applicant was compensated with a lump sum for the breach.

Even with this clarity in outlining why the applicants' constitutional rights were violated under Section 153, Chief Justice Chang avoided the substantiative issue of transgender persons being unfairly targeted and treated because of the cross-dressing ban. Instead, he cleverly engaged in hermeneutical craftiness by invoking the savings clause (Article 152(1)) to protect the dubious vagrancy law and avoid it being rendered unconstitutional. He reasoned that

> Section 153(1)(XLV11) of the Summary Jurisdiction (Offences) Act of 1893 has been part of the laws of this country since 1893 and has survived both the 1966 and the 1980 Constitutions. As existing or rather pre-existing law under both Constitutions, it has continued in force without constitutional challenge until now. (p. 24)

In upholding Section 153, coloniality is entangled in judicial interpretation as Chief Justice Chang defers to imperial posterity and the power of the legislature to amend or draft new laws instead of accepting the 1980 amendments to the Constitution that enshrine fundamental rights and freedoms that thoroughly protect the civil rights of individuals. The applicants' "main argument was that this 1983 vagrancy law, which uses terms like 'improper purpose,' 'male attire' and 'female attire,' is very vague and fails to give the person of ordinary intelligence a reasonable opportunity to know what is prohibited" (U-RAP, 2018, p. 1).

The structural hermeneutical inequality of Section 153 violates the rights of transgender persons because it gives discretionary powers to the courts and other gatekeepers to interpret violations to their liking, placing transgender persons at a disadvantage when they are already negatively judged by cisgender heteronormative standards.

> The legislature has intentionally used free-standing terms leaving it to the court to determine as a question of fact in the prevailing social conditions and circumstances whether a particular purpose is improper and whether a particular piece of attire is 'male' or 'female.' (U-RAP, 2018, p. 25)

This creates epistemic gaps that do not favour mitigating injustices for marginalised persons.

Chief Justice Chang engages in hermeneutical duplicity in justifying the law while minimising its impact on transgender persons through a kind of gender dissemblance. The Chief Justice states: "It is instructive to note that it is not a criminal offence for a male to dress in female attire and a female to wear male attire in public way or place under 153(1)(XLV11). It is only if such an act is done for *an improper purpose* that criminal liability attaches." This implies that it is not a crime to be transgender; however, if a transgender person behaves in a way deemed as an "improper use" in public, then it is a crime.

With transgender persons' right to freedom of expression undermined through Section 153 being shielded by the savings law clause, Bulkan and Robinson (2017), p. 229) rightly argue that "criminal laws such as vagrancy laws have always generated concerns because they literally encompass so many innocent acts." Transgender bodies are constructed as disruptive and disorderly, allowing transphobia and homophobia to prevail in controlling and punishing genderqueer expressions as well as restricting same-sex intimacy, given Guyana's strict anti-buggery law.

The applicants also argued that Section 153 was a form of sex and gender discrimination against transgender persons. However, since dominant notions of gender are constructed through a cisgender heteronormative lens, the recognition of equal rights for transgender persons was dismissed. Contemporary Caribbean feminist and queer scholars have challenged gender essentialism in culture, institutions, policy, law and the state, which reproduces gender binaries that maintain a strict man/woman dyad based on Enlightenment norms to the exclusion of multiple and intersecting gender identities (Barriteau, 2004).

Bulkan and Robinson (2017), as legal experts, confirm the precarity of litigating transgender rights in gender discrimination cases. They state that:

> Sex/gender discrimination is comprehended primarily in terms of a comparison between males and females. When laws disadvantage both males and females because they are grounded in stereotypes about masculinity and femininity, they have registered as neutral to courts, rather than profoundly discriminatory. (Robinson & Bulkan, p. 232)

This blind spot in the law does not allow for a queer understanding of gender, which is on a spectrum and can include a myriad of sex/gender configurations (Butler). Accordingly, Chief Justice Chang reasoned that:

> In pith and substance, Section 153(1) (XLV11) creates a prohibition against both male and female persons equally acting in the same manner for any improper purpose and does not discriminate against male or female person on the basis of gender. (p. 28)

The structural hermeneutical injustice against transgender persons is glaring here. Because they are not deemed comparable subjects to either cisgender men or cisgender women in the adjudicative process under equal treatment discourse, they are neutralised into a normative binary gender system with limited legal recourse for harm and discrimination perpetrated against them. Trotz (2013, p. 3) rightly argues that "Justice Chang's decision that fairness has to do with equal application rules out careful attention to the content of the law itself." In the end, the High Court in 2013 and the Court of Appeal in 2017 upheld the cross-dressing ban under the armour of the savings law clause, to the disappointment of the appellants. Those in power were unwilling to use the collective social imagination and hermeneutical resources to embrace an "openness to differences" (Medina, 2013, p. 266) that values and protects the dignity and rights of transgender persons.

Appeal to Caribbean Court of Justice, Transgender Rights and Epistemic Justice

On 28 June 2018, the CCJ heard the case based on the arguments previously presented by the appellants, with the savings law clause at the forefront of the debate. On 13 November 2018, the five Justices of the CCJ unanimously upheld the appeal, concluding that Section 153 was unconstitutional as it violated the rights of the appellants by discriminating against and criminalising them due to their gender identity and expression as transgender persons. The appellants were not afforded equal protection under the law due to unfair treatment. The CCJ reasoned that "Article 149(1) of the Constitution protects the people of Guyana from discrimination. No law can be enacted that is discriminatory of itself or in its effect" (p. 62). Thus, the dubious savings law clause could not shield the cross-dressing ban. Chief Justice Saunders emphasised the importance of the "dynamic" nature of the Constitution, which must adapt to changing times and be respected in its entirety rather than piecemeal when it comes to mitigating discrimination. Demonstrating hermeneutical virtue in seeking justice, he reasoned:

> If one part of the Constitution appears to run up against an individual fundamental right, then, in interpreting the Constitution as a whole, courts should place a premium on affording the citizen his/her enjoyment of the fundamental right, unless there is some overriding public interest. (CCJ, 41, 14 November 2018)

There was no escaping the supremacy of Guyana's 1980 amended Constitution in upholding the fundamental rights and freedoms aligned with regional and international human rights standards, as demonstrated by the cases discussed in this chapter. The CCJ was forthright in its defence: "The savings clause would only be needed where it proved utterly impossible to modify the existing law to make it conform with the Constitution" (p. 58). In discussing decoloniality and the law in this CCJ decision, Elliott-Williams (2020, p. 744) states "that in helping to fashion a decolonized conception of Caribbean constitutionalism, the Court is properly fulfilling its adjudicative function, in a context where legislatures have been unwilling to fulfil their law-making function."

The CCJ provided a progressive hermeneutical understanding of gender identity, moving beyond the heteropatriarchal constructs promoted by some lawmakers and the church. Instead, the Court affirmed that transgender persons are entitled to human dignity and freedom of expression, including their choice of clothing, without facing threats, harassment or harm from the law, the state or others. The CCJ stated, "A person's choice of attire is inextricably bound up with the expression of his or her gender identity, autonomy and individual liberty. How individuals choose to dress and present themselves is integral to their right to freedom of expression" (p. 76). Furthermore, in addressing the structural hermeneutical injustice in invalidating Section 153, the CCJ noted that the provision concerning "improper use" was vague and could obstruct people's understanding of what constitutes criminal action:

> The penal statute must meet certain minimum objectives if it is to pass muster as a valid law. It must provide fair notice to citizens of the prohibited conduct. It must not be vaguely worded. It must define the criminal offence with sufficient clarity that ordinary people can understand what conduct is prohibited. It should not be stated in ways that allow law enforcement officials to use subjective moral or value judgments as the basis for its enforcement. A law should not encourage arbitrary and discriminatory enforcement. (p. 80)

In effect, abuse of power arises when authorities are granted discretionary power to determine what constitutes an offence without an objective standard that is comprehensible to the majority of the populace. Epistemic injustice is perpetuated by vague and ambiguous language in law and policy, which keeps marginalised groups powerless and unaware of the actions of those in authority. The CCJ correctly ruled in favour of the appellants by invalidating the colonial vagrancy law designed to control and punish black and brown queer bodies. Moreover, the epistemic resistance demonstrated by the litigants, their lawyers, U-RAP, and the support from regional LGBTQ groups and other allies exemplifies the power of collective mobilisation and the effectiveness of strategic litigation in creating a counterpublic to defend justice.

In conclusion, dismantling hermeneutical injustice is challenging at the structural level because restrictive ideologies and beliefs are deeply embedded and normalised within systems and institutions. These systems, often dominated by privileged groups, use their influence to protect their interests and maintain their power, making it difficult to effect meaningful change.

The lack of inclusion and diversity within society undermines democracy by preventing different groups from accessing rights, knowledge and resources that could improve their conditions. LGBTQ groups and activists in the region have engaged in epistemic resistance by creating counter-publics, providing safe spaces and social outlets for LGBTQ youth, raising awareness about LGBTQ issues through consciousness-raising, human rights education and public advocacy, and through strategic litigation to challenge discriminatory laws and practices that reinforce homophobia and transphobia in law, state and church. Decolonisation in the Caribbean is an ongoing process, and the landmark cases that led to the abolishment of anti-buggery and vagrancy laws in some countries are a testament to the impact of social justice hermeneutics. They underscore the strength of Caribbean constitutionalism in safeguarding the rights of all individuals, regardless of sex, gender or sexuality.

REFERENCES

Alexander, M. J. (1991). Redrafting morality: The postcolonial state and the sexual offences bill of Trinidad and Tobago. In C. T. Mohanty, A. Russo, & L. Torres (Eds.), *Third world women and the politics of feminism* (pp. 133–152). Indiana University Press.

Alexander, M. J. (1994). Not just (any) body can be a citizen: The politics of law, sexuality and postcoloniality in Trinidad and Tobago and The Bahamas. *Feminist Review, 48*(1), 5–23.

Alexander, M. J. (2005). *Pedagogies of crossing: Meditations on feminism, sexual politics, memory and the sacred.* Duke University Press.

Amlak, W. (2013). *Beware of BGLAD's offer of help.* Nation News.

Attai, N. (2017). Let's liberate the bullers! Toronto human rights activism and implications for Caribbean strategies. *Journal of Eastern Caribbean Studies, 42*(3), 97–121.

Barriteau, E. (2004). Constructing feminist knowledge in the commonwealth Caribbean in the era of globalization. In B. Bailey & E. Leo-Rhynie (Eds.), *Gender in the 21ˢᵗ century: Caribbean perspectives, visions and possibilities* (pp. 437–465). Ian Randle Publishers.

Baker, S. J., & Lucas, K. (2017). Is it safe to bring myself to work? Understanding LGBTQ experiences of workplace dignity. *Canadian Journal of Administrative Sciences, 34*(2), 133–148.

Bowcott, O., & Wolfe-Robinson, M. (2012). Gareth Henry: I saw friends killed...there is no safe place in this country. In *The Guardian.* https://www.theguardian.com/world/2012/oct/26/jamaican-gay-petitioner-gareth-henry

Bulkan, A., & Robinson, T. (2017). Enduring sexed and gendered criminal laws in the anglophone Caribbean. *Caribbean Review of Gender Studies, 11*, 219–240.

Constitution of Belize 1981 (rev. 2011). (2011). https://www.constituteproject.org/constitution/Belize_2011

Cornell, D. L. (1992). Gender, sex and equivalent rights. In J. Butler & J. W. Scott (Eds.), *Feminists theorize the political* (pp. 280–296). Routledge.

Crawford, C. (2019). Unbearable knowledge: Sexual citizenship, homophobia and the taxonomy of ignorance. *Journal of Eastern Caribbean Studies, 44*(2), 115–144.

Elliott-Williams, G. (2019). The CCJ decolonizing Caribbean constitutionalism. *Commonwealth Law Bulletin, 45*(4), 742–751.

Fricker, M. (2007). *Epistemic injustice: Power and the ethics of knowing.* Oxford University Press.

Frontline Defenders. (n.d.). Kenita Placide testimony. Retrieved 2023, from https://www.frontlinedefenders.org/en/testimonial/kenita-placide-testimony

Gaskins, J. (2013). "Buggery" and the commonwealth Caribbean: A comparative examination of The Bahamas, Jamaica, and Trinidad and Tobago. In C. Lennox & M. Waites (Eds.), *Human rights, sexual orientation and sex identity in the commonwealth: Struggles for decriminalization and change* (pp. 429–454). University of London Press.

Ginn, D., & Kindred, K. (2017). Pluralism, autonomy and resistance: A Canadian perspective on resolving conflicts between freedom of religion and LGBTQ rights. *Religion and Human Rights, 12*(1), 1–37.

Gosine, A. (2009). The heteronationalism of MSM. In C. Barrow, M. de Bruin, & R. Carr (Eds.), *Sexuality, social exclusion and human rights: Vulnerability in the context of HIV* (pp. 95–115). Ian Randle Publishers.

Guyana's Constitution of 1980 with Amendments through 2016. (2016). https://www.constituteproject.org/constitution/Guyana_2016.pdf?lang=en

Haynes, T. (2016). Mapping Caribbean cyberfeminisms. *SX Archipelagos, 1*, 1–19.

High Court of Trinidad and Tobago. (2018). *Jason Jones v.* The Attorney General and Interested Parties.

Jade, T. (2020). *Jason Jones: The LGBTQ+ activist breaking ground in Trinidad and the United Kingdom.* Caribbean Collective. https://www.caribbeancollectivemag.com/social-impact/jason-jones

J-FLAG (2012). "Human Rights First Fact Sheet". Jamaican Forum for Lesbians, All-Sexuals, and Gays (J-FLAG). https://humanrightsfirst.org/wp-content/uploads/2022/11/Jamaica-LGBT-Fact-Sheet.pdf

Kayne, A. (2022). *Constitutional "savings clauses" revisited following the decision of the JCPC in Chandler.* Libertas Chambers. https://www.libertaschambers.com/media-hub/constitutional-savings-clauses-revisited/

Kempadoo, K. (2009). Caribbean sexuality: Mapping the field. *Caribbean Review of Gender Studies, 3*, 1–24.

Lavers, M. (2014, March 25). *Activist tells U.N. Panel LGBT people face "brutal" violence.* Washington Blade. https://www.washingtonblade.com/2014/03/25/activist-tells-u-n-panel-lgbt-people-face-brutal-violence/

Lazarus, L. (2015). Sexual citizenship and conservative Christian mobilisation in Jamaica. *Journal of Eastern Caribbean Studies, 40*(1), 109–140.

Lazarus, T. (2020). Enacting citizenship debating sex and sexuality: Christians' participation in the legal processes in Jamaica and Belize. *Commonwealth and Comparative Politics, 58*(3), 366–386.

Luton, D. (2009). Buggery laws firm—PM says life or 15 years for some sex-offence breaches. *Jamaica Gleaner.* http://jamaica-gleaner.mobi/20090304/lead/lead1.php

Mann, A. (2017, April 19). What does Barbados' prime minister have to say about the country's harsh buggery laws? *Xtra Magazine.* https://xtramagazine.com/power/what-does-barbados-prime-minister-have-to-say-about-the-countrys-harsh-buggery-laws-73370

Medina, J. (2013). *The epistemology of resistance: Gender and racial oppression, epistemic injustice, and resistant imaginations.* Oxford University Press.

Mohammed, R. (2017). B-GLAD voices: The LGBT country report. Barbados, Gays, Lesbians and All-Sexuals against Discrimination. https://www.academia.edu/36091654/VOICES_-_Barbados_LGBT_Country_Report

Murray, D. (2006). Who's right? Human rights, sexual rights and social change in Barbados. *Culture, Health & Sexuality, 8*(3), 267–281.

Murray, D. (2009). Bajan Queens, Nebulous Scenes: Sexual Diversity in Barbados. *Caribbean Review of Gender Studies, 3*, 1–20.

Reid-Smith, T. (2020). Barbados government proposes same-sex unions and hints it will make gay sex legal. *Gay Star News.* https://www.gaystarnews.com/article/barbados-government-proposes-civil-unions-and-hints-it-will-make-gay-sex-legal/

Robinson, T. (2009). Authorized sex: Same-sex sexuality and law in the Caribbean. In C. Barrow, M. de Bruin, & R. Carr (Eds.), *Sexuality, social exclusion and human rights: Vulnerability in the context of HIV* (pp. 3–22). Ian Randle Publishers.

Sexual Offences Act Barbados, CAP 154 L.R.O. (2002) Laws of Barbados https://www.barbadoslawcourts.gov.bb/assets/content/pdfs/statutes/SexualOffencesCAP154.pdf

Sheller, M. (2012). *Citizenship from below: Erotic agency and Caribbean freedom.* Duke University Press.

Supreme Court of Appeal Belize. (2016). *Claim No. 668 of 2010. Caleb Orozco v. The Attorney General of Belize.*

Trotz, D. A. (2013). The constitutional challenge to the cross-dressing law. In *Stabroek News.* https://www.stabroeknews.com/2013/09/23/features/in-the-diaspora/the-constitutional-challenge-to-the-cross-dressing-law/

Tuana, N. (2006, Summer). The speculum of ignorance: The women's health movement and epistemologies of ignorance. *Hypatia, 21*(3), 1–19.

U-RAP (University of the West Indies Rights Advocacy Project) (2018). On the McEwan and Others v Attorney General of Guyana case. The upcoming appeal to the CCJ to be heard on June 28, 2018. https://justicediversitytt.wordpress.com/wp-content/uploads/2018/06/mcewanccj.pdf

Williams, C. C., Forbes, J. R., Placide, K., & Nicol, N. (2020). Religion, hate, love, and advocacy for LGBT human rights in Saint Lucia. *Sexuality Research and Social Policy, 17*(4), 729–740.

Williams-Sambrano, M. (2018, July 20). Trinidad and Tobago's religious leaders call on government to uphold anti-LGBT laws. *Religious News Service.* https://religionnews.com/2018/07/20/trinidad-and-tobagos-religious-leaders-call-on-government-to-uphold-anti-lgbt-laws/

"It's a Girl Thing": Problematising Female Sexuality, Gender and Lesbophobia in Caribbean Culture

Abstract In this chapter, I will further complicate the discourse of sexual rights by interrogating how different subordinated groups have their own power struggles when they do not incorporate an intersectional liberatory praxis to deal with interlocking identities and oppressions of gender and sexuality (sexism and heterosexism), for example, when it comes to the oppression of lesbians or queer women whether in mainstream Caribbean feminism and/or gay activism. I interrogate how the experiences of lesbian and gender non-conforming women get overlooked in mainstream gay and feminist theorising. Essentialist notions of gender and masculinist notions of sexuality, which may appear as mutually exclusive, tend to stifle the intersectional identities and existential experiences of lesbians due to lesbophobia. I analyse mainstream perceptions about lesbians in the print media that vilify them and reinforce their subordinate status as sexual citizens.

Keywords Female homosexuality • Lesbianism—zami, mati, WSW—in the Caribbean • Heteropatriarchy • Gender, respectability and sexual morality • Lesbophobia • Representations of lesbianism in popular media

C. Crawford, *Gender, Sexual Citizenship and Epistemic Injustice in the Caribbean*, https://doi.org/10.1007/978-3-031-83493-6_5

THE SUBJECT OF LESBIANISM IN SCHOOLS is a cause for grave concern, as it has a negative effect on every level of society. It is imperative that this matter receives the attention that it warrants, so as to bring about some form of resolve to save our young people from the moral decadence that this lifestyle brings. (report by Harewood, 2010d, p. 11A)

Those of us who stand outside the circle of this society's definition of acceptable women; those of us who have been forged in the crucibles of difference—those of us who are poor, who are lesbians, who are Black, who are older—know that *survival is not an academic skill*. It is learning how to stand alone, unpopular and sometimes reviled, and how to make common cause with those others identified as outside the structures in order to define and seek a world in which we can all flourish. (Lorde, 1984, p. 112)

Lesbianism has been a cause for public concern in the Barbadian popular imagination in the last few years. Led by religious conservatives and their supporters, the first sign of discontentment was highlighted in the *Nation* newspaper's sensationalist coverage of lesbianism that took place over three consecutive Sundays in April 2010. Teenage girls were condemned for frolicking with one another, lesbians, or women who have sex with women (WSW), were asked to repent and convert back to heterosexuality, and lesbianism was equated with many social ills in society. Social angst about female homosexuality did not quickly abate because, in February 2011, the Hollywood movie *Black Swan* was temporarily banned from cinemas because of its "lesbian" content. The "L" word was being used regularly in media as if it was common practice. But this had nothing to do with a change in attitude towards homosexuality; it was, more so, a master technique based on how dominant power relations are employed to first name, and then deal with, the "undesirable" thing instead of ignoring it altogether. So, in this case, the approach of de-silencing was purposeful in simultaneously denouncing and de-legitimising same-sex female sexuality. This attack on lesbians was a clear case of lesbophobia in Barbadian society.

Like any other "phobia" which has some categorisation of aversion attached to it lesbophobia can be defined simply as the fear, dislike or hatred of lesbians or women who are sexually, physically and/or emotionally attracted to other women whether on an individual or group level. But scholars such as Kulich critique the use of terms that have 'phobia' attached to them, like "homophobia" and "lesbophobia," because there is a tendency to suggest that perpetrators of bigotry have some kind of socio-psychological problem that does not make them fully responsible for their feelings of panic and/or actions of contempt towards "nonnormative sexualities and genders" (2009, p. 24). As a result of this, the litigious behaviour of homophobes is often reduced to an individual pathology instead of being linked to the structural heteronormative codes. Despite this, I think there is a political importance and relevance in using the term homophobia, and lesbophobia specifically, because attitudes of disdain (more so than fear) and violent actions against LGBTQ people do occur and are debilitating to individuals who are doubly victimised because of anti-buggery laws in most Caribbean countries. The term "anti-gay"does not capture the same intent to hurt, harm and exclude non-heterosexuals in society. I heard a woman say in an academic setting that she is "anti-gay"—or heterosexist—but not homophobic. She, in turn, takes a sort of moral high ground on the issue because she is not exhibiting aggressive behaviour towards gay men and women. In this case, erratic homophobic behaviour is substituted by the liberal stance of tolerance: *I don't accept you, but I will put up with you if you don't get in my way.* But ultimately, heterosexist ideology legitimises homophobic acts, whether this is in the form of harassment, discrimination and/or violence.

Lesbophobia culminates through the intersection of sexism and homophobia as two mutually constituted regimes of oppression that produce the effects of harm—whether this is prejudice, harassment, discrimination and/or sexual and physical violence—against women who love and have sex with other women. Capezza (2007) notes that sexism and homophobia are embedded in traditional gender role identification and expectations for men and women. She goes on to argue that.

> Traditional gender role beliefs are linked to sexism and in turn to homophobia due to perceived violations of traditional gender role expectations. If a person endorses such traditional gender role beliefs, then they are [more] likely to express hostility toward individuals who violate these norms, such as nontraditional women (e.g., career women) or homosexual men. (2007, p. 249)

While gender ideology shapes and normalises men's and women's perceptions and attitudes about masculinity and femininity and produces asymmetrical power relations between men and women (Barriteau, 1998), there is a more substantive ideological basis to lesbophobia that gives it weight and legitimacy. Drawing on Jacqui Alexander's (1991) work on female sexuality, morality and state control, I argue that lesbophobia is the byproduct, or an effect, of a heteropatriarchal ordering of gender and sexuality that simultaneously privileges and reinforces heterosexuality or opposite sex relations (heterosexism) and men's dominant (patriarchal) claims over women's bodies for physical, sexual and reproductive purposes. Atluri adds that "both lesbians and gays threaten the *natural, moral* state of heterosexual, patriarchal family, and therefore their suppression is often integral to maintenance of patriarchy" (2001, p. 12). Therefore, the individual and institutional efforts to police and control lesbians are proscriptive in restricting female sexual autonomy that is freely expressed, not solely procreative, and that may not involve or focus on men.

In this chapter, I will examine how lesbophobia manifests in a postcolonial Caribbean landscape in multiple ways, whether it is through societal sanctions such as stigma, discrimination and violence, or through fabricated claims of sexual immorality against same-sex female sexuality promoted by the church, state and media. From a critical feminist perspective, I will first critique dominant notions of gender and sexuality by exploring the relationship between patriarchy and heterosexism in ordering female sexuality and sexual relations. I will then discuss the ways in which lesbian sexuality and bodies are constructed to denote a kind of corporeal disorder that is unsettling or disruptive to dominant notions of hetero-femininity or womanhood associated with gender identity, sexual pleasure and motherhood. Finally, I will demonstrate how the media plays a role in manufacturing and reinforcing lesbophobia through sensationalist accounts that serve to pathologise and delegitimise same-sex female sexuality.

Feminism, Male Homosexuality and the Obscure Lesbian Subject

How has lesbianism or female same-sex sexual relations been explored and located within, and across, Caribbean culture? Except for Silvera (1992), Alexander (1991, 1997, 2005), Elwin (1997), King (2008), French and Cave (1995), Wekker (1997; 2006), Tinsley (2011) and the anthology by Glave (2008) that captures both gay and lesbian subjectivities and experiences through fiction and non-fiction writing, there is a paucity of scholarly research that has *thoroughly investigated or theorised female homosexuality* in the Caribbean beyond a cursory glance. Documentation of the diversity of female same-sex sexual experiences in the Caribbean, across race/ethnicity, class and culture, is even more scant. Furthermore, at times the gendered-sexualised subjectivities of lesbians tend to get subsumed, or overlooked altogether, when discussing women (*read as heterosexual*) and gay men, generally, as subordinate groups within a heteropatriarchal order. This *homogeneity of difference*, which Lorde (1984) cautions us about, is just as troubling as *intolerance to difference* based on essentialist notions of gender and sexuality and monolithic constructions of collective identity.

I think that there is a particularity, and a peculiarity, in the ways in which lesbians are marginalised in society. The subordination that lesbians face is clearly borne by them violating, or maybe more discursively transgressing, dominant norms of gender and sexuality. But the *peculiar* aspect of the subordination of lesbians is somewhat more nuanced in understanding based on their intersectional identity and "nomadic" existence and movements between different social locations and categories, such as "Woman" and "Homosexual" (Braidotti, 1994). Caribbean feminists have made valuable contributions to examining women's subordination to men in relation to how asymmetrical gender relations operate through the sexual division of labour via family, work and political economy and through exclusionary practices of the church and state to disadvantage women and confer more rights, power and privilege to men than to women (Barriteau, 2003, 2004; Massiah, 2004; Mohammed, 2002; Reddock, 1994; Robinson, 2003). Other scholars have looked at violence against women (Clarke, 1997), female sexual vulnerability and HIV/AIDS (Roberts et al., 2012; Murray, 2009), and women's participation in commodified sex markets, such as sex tourism and prostitution (Cabezas, 2004; Kempadoo, 2003). But the heterogeneity and complexity of women's gendered identities and sexual relations must be more thoroughly investigated beyond a heteronormative lens. Men's relationship to, and with, women tend to be taken as a given here both socially and sexually. It is rarely questioned how lesbians, in defying codes of heterosexual femininity, may have less leverage in negotiating power relations with men on a personal and public level. In addition to this, some liberal feminists, in their quest for equality with men, may overlook how their own heterosexual privilege in women's organising and civil liberties does not take into consideration how the rights of lesbians are denied (such as in marriage and adoption, domestic violence laws that exclude same-sex couples and laws that criminalise sex between women).

Scholars have also investigated homosexuality in the Caribbean focusing on homosexual male experiences and non-normative gender and sexual expressions and sanctions against them by their families, church and state (Crichlow, 2004; Glave, 2008; Murray, 2009). There has also been an examination of hegemonic masculinity in shaping dominant heterosexual male norms on gender and sexuality that contribute to hypermasculinity and homophobic sentiments (Chin, 1997; De Moya, 2004; Lewis, 2003). Finally, there have been discussions about stigma and discrimination against MSM and the challenge in combating HIV/AIDS (Carr, 2005). While these perspectives are instructive in highlighting how homosexual men are constantly being threatened and surveilled in society (inclusive of acts of public violence used against them) for deviating from hegemonic codes of masculinity, the category "homosexual" seems to uphold androcentrism which privileges masculinist perspectives of same-sex desire and does not address the misogyny that might be produced against lesbians, and women generally.

FEMALE SEXUALITY AND THE LESBIAN THREAT

Davina Cooper, in *Power in Struggle: Feminism, Sexuality and the State*, states that "as a form of disciplinary power, sexuality organizes identity, knowledge, behaviour, manners, dress and social interactions around particular desires, libidinal practices and social relations" (1995, p. 67). Enlightenment constructions of femininity, as docile, pious, chaste and procreative, have rendered sex for women as a functional act, not for pleasure, and dependent on men. Male sexuality is recognised as a force to be reckoned with, powerful, expansive and penetrative whereas female sexuality is seen as a passive and receptive force of the latter. Kitzinger argues that "sex, as it has been constructed under heteropatriarchy, seems necessarily to involve the eroticizing of power and powerlessness, dominance and subordination: that is what makes it erotic" (Kitzinger, 1994, 207). In this case, women's sexuality is restricted through men's claims over their bodies for sex and reproduction; and heterosexual sex is simply reduced to missionary position—man on top and woman on the bottom. While men have been granted sexual autonomy through codes of hegemonic masculinity, women have been seen as relying on men for sex, and *only* receiving it through them. Due to sexual double standards and codes of morality, women, unlike men, who are sexually free and uninhibited, in wanting and demanding sex, are often ridiculed and characterised as "whores" and 'sluts' or loose women because they violate gender and sexual codes of "respectable" femininity.

In the Caribbean, lesbians or women who do not conform to hetero-sexuality as a compulsory or standard way of life, or those women who challenge rigid gender codes of femininity, sell their sex for money or who do not adhere to heterosexual monogamy are viewed as disruptive to the dominant heteropatriarchal order. Rosamond King points out that respect-able femininity in the Caribbean has been informed by power relations in terms of race, class and cultural lines from our colonial past. She states, "black and brown Caribbean women's sexualities have always been con-sidered 'queer,' odd, and less moral by European (and often by 'coloured') elites. Women who choose extramarital sex and childbearing, non-monogamous relationships, non-nuclear family structures, or lesbianism have always been maligned by those in power" (King, p. 193).

In examining female sexual morality, state and the law in Trinidad and Tobago and the Bahamas, Jacqui Alexander (1991, 1997) points out that Caribbean states in the post-independence period have adopted master techniques through legislature derived from European colonialists to police and regulate gender and sexual relations. In examining the Sexual Offences Bill (1986) for Trinidad, Alexander goes on to argue that moral-ity and economics converge in the law whereby sexual relationships that do not reproduce a workforce are seen as morally corrupting; therefore, they are deemed illegitimate and deserving of punishment. Heterosexual procreative relations, especially within marriage, were upheld through the continued criminalisation of anal sex (buggery) between men and men (and between men and women), as "unnatural" and indecent, along with the criminalisation of same-sex relations between women. Alexander states that:

> Biology and procreation sanction nature and morality to such an extent that when eroticized violence threatens to dissolve heterosexual conjugal mar-riage, a textual restoration is enacted by criminalizing lesbian sex and sex among gay men—an act of reasserting the conjugal bed. Indeed, the rein-scription of the conjugal bed occurs precisely because no alternative sexuali-ties are permissible; by legally outlawing other alternatives that 'reject the obligation of coitus,' the power of marriage is reinscribed, and with it the reinforcement of 'obligatory social relationship between "man" and "woman"'. (1991, p. 138)

Lesbians are seen as particularly threatening to the status quo because they outrightly challenge compulsory heterosexuality—the idea that women should be, and want to be, with men. Adrienne Rich states that "lesbian [or same-sex female] existence comprises both the breaking of a taboo and the rejection of a compulsory way of life. It is also a direct or indirect attack on male right of access to women" (1993, p. 238). Because so much of economic and socio-cultural operations, whether productive or reproductive, rely on the myth of femininity for the purposes of capital accumulation, caring for others and male aggrandisement, lesbians challenge male authority because their work and bodies cannot be readily tapped into on a private level. Sexism underlines the "man-hating" indictments directed against lesbians since women generally are not expected to be sexually engaged, and powerful, without men. Lesbianism exposes the unstableness of heterosexuality. Butler argues that "for, if to identify as a woman is not necessarily to desire a man, and if to desire a woman does not necessarily signal the constituting presence of a masculine identification, whatever that is, then the heterosexual matrix proves to be an *imaginary* logic that insistently issues forth its own unmanageability" (1993, p. 239).

This also raises questions about how lesbians are positioned in relation to motherhood and family. Dominant Euro-American norms of gender and sexuality have defined what motherhood *is*, and *is not*, based on a nuclear heterosexual family model and exclusive mother–child relations (Crawford, 2011). The institution of motherhood (Rich, 1995) is premised on heteropatriarchal relations: "Motherhood is what mothers and babies signify to men" (Rothman, 1989, p. 27) so women have children with, and for, men (Comeau, 1999). Therefore, lesbians are not seen as legitimate mothers in the areas of gender, sexuality, reproduction and familial relations (Benkov, 1998) because men do not privately control their sexual and reproductive labour. Although in the Caribbean there is a high visibility of female-headed households and matrifocality validates the central role that elder women play in caring for children and others (Barrow, 1996; Clarke, 1999; Mohammed, 1998; Smith, 1996), there has been little investigation or discussion of how some of these childcare and familial arrangements occur for lesbians, outside of a heterosexual and/or Euro-American nuclear family norm. While women, generally, tend to be problematically de-sexualised as mothers in reinforcing codes of female morality and chastity (*you cannot be sexual and maternal*), I think that for lesbian mothers the opposite is true. The de-sexualisation does not readily occur due to stigma against lesbian sexuality as morally corrupting. The "good mother" construct relies on women conforming to codes of respectable hetero-femininity. Since there is greater threat of lesbian mothers being seen as unfit mothers or "bad mothers" due to their non-heterosexual lifestyle, many women may lead closeted lives to protect themselves and children from ridicule and discrimination or to prevent losing their children in custody cases (Benkov, 1998). While in some incidences heterosexual women may be valorised for their role as mothers, this praise or privilege is not readily extended to lesbian mothers.

Delegitimising Lesbian Sex

Since traditional research on human sexuality has been informed by androcentrism and phallocentrism (Williams & Stein, 2002), in the heteropatriarchal imagination, lesbian sex tends to be rendered not *real sex* because of the absence of a penis. There is curiosity about what two women do sexually since in popular lore sexual activity is reduced to intercourse or coitus whereby the male sexual organ is objectified and seen as dominant. Thus, lesbian sex is de-legitimised as non-sex because women are perceived as needing men to satisfy them sexually. While lesbian libidinal desires vary, with women pleasuring women in different ways inclusive of penetrative sex, in popular discourse lesbian sex is either passive or vanilla of sorts or involves pornographic scenarios of two ultra-feminine women (usually straight) engaging in sexual play, produced by, and consumed through, the male gaze, symbolically phallic, for the pleasure of heterosexual men. Hegemonic masculinity thrives in reproducing and maintaining gender and heterosexual conformity. In interrogating masculinity in the Caribbean, Linden Lewis explains:

> Hegemonic masculinity refers to practices of cultural domination of a particular representation of men and manliness. It refers to an orientation that is heterosexual and decidedly homophobic. It prides itself on its capacity for sexual conquest and ridicules men who define their sexuality in different terms. Hegemonic masculinity often embraces certain misogynist tendencies in which women are considered inferior. Departure from this form of masculinity could result in a questioning of one's manhood. (Lewis, 2003, p. 108)

Although gay male sex is abhorred by homophobic heterosexual men as being unnatural, there is a way in which heterosexual men view gay sex as involving *real* sexual activity due to the corporeal phallic threat to their masculinity. Since dominant notions of sexuality position heterosexual relations as *natural* or *normal* sexual activity based on gender binaries— with men dominating or "doing it" to women as the *active* masculine over the *receptive* feminine—then two men having sex together tends to be viewed through a heterosexist lens (Butler, 2007). One male partner is perceived as masculine and dominant and the other as feminine and subordinate. Thus, homophobia and sexism work in tandem in preserving heterosexual masculine power through sex/gender dichotomies (Richardson, 1998). Feelings of emasculation by heterosexual men stem from fears of being sexually propositioned by homosexual men based on stereotypes about male homosexuality. Lesbians, on the other hand, as women, are not viewed as physically or sexually threatening to heterosexual men.

Gender Ambiguity and a Queer Lesbian Identity

Lesbians, and women in general, who break gender codes by not being clothed in representations of "femininity" or who have more masculinised features or appearances— androgynous, tomboy or butch—are seen as aberrations to the normative gender regime. Gender non-conforming lesbians are contemptuously characterised as "hard," "man-like," "man royals," "bulldagger" and the like. A queer lesbian identity clearly violates the cult of femininity in both bodily performance and behaviour but it is also unsettling to hegemonic constructions of masculinity that classify the androcentric subject as being solely a biological male (Butler, 2007). Gender ambiguity (genderqueer or transgender) in lesbianism that is noticeable is often translated into intolerance and violence against women because they defy codes of hetero-femininity. Both gender identity and sexual identity are called into question. Since clothing is also important in how gender is performed, one's gender identity is often conflated with sexual orientation when an individual's appearance seems to deviate from "appropriate" representations along the masculine–feminine scale. Generally, a woman might be held suspect of being a lesbian if she does not wear stereotypical feminine attire (wearing dresses, high heels, make-up, etc.) or behave in a gender-specific way, even if she is not gay (Silvera, 1992). Butch lesbians are seen as particularly dangerous to the sex/gender dualistic order, which relies on mutually exclusive categories, because of their gender non-conforming expressions and behaviours. Likewise, female athletes are targets of lesbophobic sentiments, regardless of their sexual orientation, because their corporeal stature contravenes strict gender assignments. Moreover, gender ambiguity in lesbianism promotes a queer lesbian identity that contravenes strict categorisations based on sex, gender identification and desire. It offers an alternative way of conceptualising and understanding how the female body can be marked by different gender and sexual identifications, as multiple and malleable, beyond essentialist ways of being. Therefore, it is important to further investigate how power operates, and is exercised, in same-sex female relationships given their variation.

In this section, I discussed how heteropatriarchal ideologies are instructive in delegitimising lesbians as women based on dominant notions of gender and sexuality. Lesbophobia is a byproduct of this and is further manifested on the practical level through the interplay of sexism and homophobia; therefore, lesbians are devalued and discriminated against because of their gender as well as their sexual orientation. The specific form of oppression that lesbians encounter because of lesbophobia will be discussed in the next section.

LESBOPHOBIA: DIRTINESS AND DISORDER

Violence against Lesbians

Gail Mason (2002) examines violence against lesbians as homophobic and anti-lesbian acts. She emphasises that both gender and sexuality inform the particularised violence against lesbians. While Mason credits feminists for taking a strong stance against male violence against women, especially in intimate partner heterosexual relationships, through activism, advocacy and legislation, she argues that there is a paucity of feminist literature when it comes to the "specific problem of homophobia-related violence towards lesbians" (2002, 38). Similarly, literature on homophobic violence tends to focus on gay male victimisation. While gay men and lesbians are targets in public spaces, with gay men being particularly vulnerable to random violent acts against them on the streets, lesbians encounter additional aggravation in personal and private situations. Mason suggests that for lesbians "a significant proportion of incidents take place at home or work, involve on-going campaigns of harassment, and are committed by one, older man acting alone, who may be known to the woman" (2002,

38). Furthermore, the sexualised-gendered violence against homosexual women because they are "lesbians"—in fact, hate crimes—includes physical and sexual assault from beatings, sexual molestation, rape (both individual and gang related) and/or sodomy.

Male power, desire and violence coalesce as lesbians are sexualised, demonised and then, ultimately, punished for their gender and sexual non-conformity. While some heterosexual men might sexually harass lesbians in similar ways to other women based on gender—due to (hetero) sexist beliefs and attitudes that reinforce men's claims to women's bodies determining this as a patriarchal right—there is another dimension to their abusive behaviour because of homophobic attitudes. There is both attraction and repulsion when a woman's lesbianism is uncovered. There is the heightened excitement that men derive from conquering a doubly unavailable female source while at the same time men may harbour feelings of disdain towards lesbians because their sexual disinterest in men is taken as a personal attack or a rejection of their masculinity, which is defined through heterosexual acts (Mason, 2002). Moreover, the attempt by men to "fix lesbians" by having forced sexual relations with them is indicative of how men will use violence to reinforce male dominance and legitimise hetero sex. Lesbians who are identifiably gay have been prime targets for lesbophobic acts against them in the form of gang rape in Jamaica (Williams, 2000).

Throughout the Caribbean LGBTQ groups have been vigilant in denouncing homophobic acts that have led to stigma, discrimination and violence against LGBTQ individuals and those presumed to be LGBTQ. They have advocated on various levels to ensure social justice for LGBTQ people both in relation to civil liberties and human rights. Democracy is curtailed by homophobic beliefs steeped in fundamentalist religious moralism that privilege heteropatriarchal theocracy over rights in *defining* and *deciding* who is worthy of equal and fair treatment in society. LGBTQ people in the Caribbean are constantly negotiating their identities and realities within a heteronormative landscape. While many are contributing to the growth and development of their societies, and carving spaces to socially convene and establish community linkages, the politics of exclusion through homophobia—from isolation and ostracism from family and friends, slurs and epithets in everyday life, being mocked, stalked and threatened, being denied services and protection before the law to sexual and physical violence—operate to control and police LGBTQ people, keeping them in a state of fear and self-surveillance.

There is a public/private division related to the way in which homo-phobic violence manifests itself differently for gay men than for lesbians. While gay men are assaulted in public usually in mob style or in front of a crowd to shame and punish them, there is a private dimension to how violence takes place against lesbians, which makes it seem less apparent and less visible (Mason, 2002). Lesbians are more vulnerable to attacks by men in their private and community spaces and the assaults tend to include physical and sexual violence, and sometimes mutilation of the genitalia (Du Long, 2005). The perpetrators usually know the women and/or they are familiar with their whereabouts. For example, in Jamaica in 2006, two women who lived together were found murdered. It was alleged they were in a relationship and lesbian content was found on the scene. "Police quickly named an estranged male partner of [one of the victims) as the prime suspect and said the apparent relationship between the women was the likely motive for the crime" (Human Rights Watch, 2006). In another account, a woman was gang raped and then murdered in her community after some men found that she was a lesbian. They did not want her to spread her "disease" to the rest of the women in the community (Du Long, 2005). An LBT women's group in Jamaica, called Women for Women, notes on their website that lesbophobic attacks are underre-ported. Because of the anti-sodomy laws, lesbians may be less likely to come forward with cases of rape and other forms of sexual assault because they fear further abuse and persecution by law enforcers and the state (WFW, 2010).

Lesbians as Disorderly Subjects: Dirtiness and Contamination

Lesbophobic sentiments are always reinforced through lesbianism being seen as a corporeal "disorder," which is signified through "dirt' and "con-tamination." Mason points out that "dirt" or "dirtiness" or what is believed to be unclean has long been associated with both homosexuality and women's bodies. If what lesbians do sexually, as LGBTQ people, is deemed unnatural or a disease, and the dominant order is in turn repulsed by it, then discrimination and violent acts against them are seen as justifi-able. The "dirtiness" of lesbians as "disorderly subjects" is also expressed through misogynist beliefs about women's vaginas (Mason, 2002, p. 46). In popular lore, women's vaginas have been equated with uncleanliness and pollution, whether through menstruation or childbirth, where fluids and odours are emitted (salty, fishy, musky). But there is also a heightened fear of dirtiness—and of contamination—in imagining two women engaging in tribadism (two vaginas rubbing together). So, lesbophobia is expressed and operates on many different levels, even on a linguistic basis: "The language of dirt functions as an effective insult because it invokes corporeally specific images of lesbian sexuality" (Mason, 2002, p. 47).

The notion of lesbianism being "dirty" or "nasty" is captured in Atluri's work on homophobia, heterosexism and nationalism in the Commonwealth Caribbean. She recounts a discussion that ensued on the walls of one of the female bathrooms at a university campus because of an ad or request being posted that read: "Want pussy to suck email me at [...]" (Atluri, 2001, p. 18). Someone responded with utter disdain and wrote back:

> Re: To the slut who wrote the above and any other lesbian garbage on campus. With so many men out there how, the hell could you even dream of wanting a wanting a woman! There's absolutely nothing remotely sexy about a woman. Lesbianism is pure nastiness and wutlessness. Gun shot to you all. Yours Sincerely, A REAL woman! (Atluri, 2001, p. 18)

Lesbophobia operates in different ways in this scenario. In the first instance, the rebuke against lesbians based on washroom graffiti is telling of how lesbians violate dominant standards of womanhood in the respondent's eyes due to gender and sexuality. In upholding heterosexism and patriarchal sex/gender relations, lesbian sexuality is read as deviant because "REAL" women are sexually attracted to men, and they should ultimately desire men and NOT women. As disorderly subjects, the body and sexuality of the lesbian woman are marked as *dirty* on two counts, in turn, contravening respectable hetero-femininity: lesbian sex is seen as corporeally unclean or "pure nastiness," and lesbian sexual behaviour is denoted as "wutlessness" (promiscuity or looseness). Terms like "slut," "bitch" and "whore" were further used to insult the person who wrote the salacious ad/request. Finally, homophobic violence is symbolically evoked against lesbians, to "right" a "wrong" behaviour, through the sentiment: "Gun shot to you all."

LESBOPHOBIA IN THE BARBADIAN POPULAR MEDIA

Same-sex relationships between females at secondary schools across the island [Barbados] are causing authority's great concern. According to reports, the problem has gotten so out-of-hand during the past two to three years that some principals and teachers have had to find ways to protect first and second form school students from being pounced upon by older students who seek to recruit them into their circles. (Harewood, 2010a, p. 5A)

The *Nation* newspaper's coverage of lesbianism in Barbadian society, which took place over three consecutive Sundays in April 2010, demonstrates how lesbophobic beliefs operate to pathologise same-sex female relations. In this case, patriarchal religious ideologies collude with the media to reinforce heteronormative moralising ideals about female sexuality, dismissing the variation of women's sexed lives that are not exclusively heterosexual. As disorderly subjects, lesbians are presented as deviant and morally corrupting to women and ultimately a threat to the family and to straight men. A woman named Sherry-Ann stated in the Week Two coverage that: "I know a lot of men who do not mind having a lesbian for kicks, but they are now disgusted because the women are taking away their women" (Harewood, 2010c, p. 13A). In this case, the thought of lesbian sexuality as a legitimate sexual preference outside of masculine persuasion raises concern because the heteropatriarchal order is doubly threatened— men do not have access to these women and lesbians might be sexual competition for men. Mason makes an important point in relation to how heterosexism operates on an ideological level: "As a discourse, the straight mind does not see lesbian sexuality as a legitimate sexual preference with a value of its own. Rather, lesbianism represents the rejection of a social order, which decrees that only men should be entitled to sexually exchange women" (Mason, 2002, p. 50). Moreover, in the coverage there is a major stake in keeping all women in their place. Patriarchal religiosity is invoked to scare teenage girls into compliance. A woman named Nicole warned: "Young people must be made to know that God does not want us to experiment" (Harewood, 2010c, p. 12A).

Lesbians are, unequivocally, presented as disorderly subjects in the *Nation*'s tri-Sunday coverage of lesbianism. Lesbians are seen and presented as both deviant and dangerous to readers to manufacture lesbianism as a social problem that needs to be fixed for the good of the public. The misapplication of utilitarian principles to denounce lesbians, through the print media, demonstrates how the systemic nature of lesbophobia is produced and reproduced in a public forum. The "Lesbian Problem" is summed up in the following points:

- The fear of contamination is invoked as girls are warned to stay away from lesbians and homosexual activity in general. Since there is the possibility that anyone can engage in homosexual acts, there is the fear of sexual boundaries being violated. Repression is needed to prevent any hetero–homo crossovers. This inadvertently speaks of the malleability, or the not fixity, of sexuality, although it was not intended by the informants; and, ironically, it challenges the so-called naturalness of heterosexuality.
- The deviance and the dirtiness of lesbians are promoted through lesbophobic sentiments. Lesbianism is not a "normal" sexual behaviour or is reduced to a "lifestyle" and is ridiculed through religious edict: "woman was made for man."
- Lesbianism is a dysfunction that is brought on by abuse, sexual coercion or familial breakdown.
- Lesbians are sexual predators: they are sexually promiscuous and are out to get or recruit teenage girls.
- Cultural relativism: lesbianism is not accepted in the Caribbean; it is just tolerated. Influences from outside (Hollywood) are leading girls astray with this kind of lifestyle.
- Identity obscurity: displays of same-sex female relationships are reduced to a lesbian identity, without fully knowing what girls are feeling and how they identify.
- Sexual repression: teenage girls should avoid same-sex sexual experimentation.
- Women can be saved from lesbianism if they repent and accept God in their life, redeeming them as respectable heterosexual women.

Master techniques via the print media are employed through sensationalist, anecdotal accounts to highlight the threat of the "lesbian menace." This biased perspective is explicitly and unapologetically lesbophobic. The coverage began on Sunday April 11, 2010, with the personal accounts of Marcia Weekes, counsellor, playwright and founder of Praise Academy, who claims the incidences of lesbianism in schools are on the rise and attempts should be made to stop such behaviour (religious influence). Her concerns are expressed as follows:

> The growth of bisexual and lesbian relationships in Barbados, and even the wider Caribbean, has escalated in the past two years, according to counsellor Maria Weekes. And she is deeply worried. (Harewood, 2010b, p. 14A)

Weekes' claim of the increase of lesbianism is speculative yet she presents it as factual to align with her biases to heighten fears about homosexual women who she feels need to be controlled. How and where would lesbians be recruited? And can all girls/women who engage in sexual activity with other girls/women be classified as lesbians? It is obvious that the motivation to quantify "lesbianism," in this case, is based on the presumption that its occurrence is something out of the ordinary, outside of the heterosexual norm. But I think that lesbian *existence* and *occurrence* are not one and the same here. Weekes is not questioning lesbian existence— she has seen *it* or has come to know *it* through "othering" sexual difference—but she is, instead, calling to attention the rate of, or propensity for, lesbianism. Fear is incited based on the possibility of mutation because of contamination via the spread of "dirt" conceived through the act of lesbianism. The warning is sounded: We will tolerate a few of you but not too many.

> Weekes goes on to state: "Young female couples are seen at times displaying their love for each other in public spaces like Queen's Park (a popular meeting place), at the beach, on the street corner – even in the school corridor and the classroom" (Harewood, 2010b, p. 14A)

The agency that girls are displaying to the public challenges hetero-norms and the assumption that homoerotic displays and desires should be contained to the private sphere. But for Weekes the closet is being opened too wide, which is contributing to the so-called braziness or boldness of girls who are disrupting standards of respectable hetero-femininity. In fact, lesbian invisibility (what is hidden from public view) is required to make sure that compulsory heterosexuality is maintained for women. Girls could not possibly be genuinely attracted to other girls, because they are supposed to *naturally* like boys, so instead something perverse is taking place. Weekes then attributes lesbianism to several factors such as vice, abuse, personal problems and familial breakdown. Her lesbophobic rant is venomous and hypocritical because she does not seem too concerned about the morals of girls being corrupted by boys who might be visibly groping or rubbing up on girls or having sex with them in deserted public spaces.

Clearly, the lesbophobic sentiments in the coverage are purposeful in heightening fear in individuals by conveniently, and dangerously, promoting bigotry through a self-professed moral authority that seeks to protect the public from sexual indecency. As Weekes professes: "I was at a particular school telling a group of females that lesbianism is wrong" (Harewood, 2010b, p. 15A). Therefore, homophobes who believe homosexuality is a sin think that they have the right to impose their ideas on others because heteronormative structures allow it. Hence, moralism trumps rights when discussing sexual minorities in the Caribbean. Social justice is obscured by a parochial belief system.

Weekes paternalistically seeks to counsels those who have fallen: "It's very strong in the arts, but I make it clear from a leadership standpoint that if a person has an issue with their sexuality, we will do whatever we can to help. No person should feel comfortable living that kind of lifestyle" (Harewood, 2010b, p. 15A). Homosexuality gets reduced as a "lifestyle" as a part of a fad subculture that is whimsical, transient and unstable, unlike heterosexuality, which is not read as a lifestyle in and of itself. This concern about a homosexual lifestyle is also voiced in Week Two's coverage:

'Many people think Barbados is a sheltered society, but a lot of ordinary-looking men and women are into this lifestyle.'

'It's all over the island today, especially in the schools. Some hide out in churches, and some

are paid [as a means of living] to engage in same-sex relations.' (April 18, 2010, pp. 12A–13A)

So, the solution to the surge in a homosexual "lifestyle" is conversion to heterosexuality through the help of the church. Being saved and further indoctrination is the prescription to getting women back on track in becoming dutiful wives and mothers, which lesbianism supposedly threatens. The issue of sexual conversion brings up the idea of malleability of sexuality. Ironically, if you can change from homo to hetero then the other way is also possible, in turn, contesting the naturalness of heterosexuality as proposed by Weekes and her supporters. But espousing lesbophobic beliefs is necessary in policing female sexuality and preventing hetero-to-homo crossovers.

Lesbianism is also pathologised through it being seen as a byproduct of a disorder or a dysfunction caused by family breakdown, low self-esteem, abuse or sexual coercion. It is not seen as a legitimate form of female sexuality whereby young women seek pleasure and intimacy from other young women just because they find it desirable. Weekes states that:

They are looking for unconditional love at home; and because many are not getting this kind of love, they are acting out in different ways. Some are young people who were violated from as early as five or six years old; so they experiment, even from primary school levels, with one another. (Harewood, 2010b, p. 15A)

The causal link between lesbianism and maladaptive behaviour and/or social malaise is faulty. Weekes overlooks the fact that many girls who are abused or who are facing familial and personal challenges are not lesbians, nor are they drawn into lesbianism. Trying to find the *cause* of lesbianism suggests that what girls are doing is out of the ordinary and is not a part of teenage sexuality; heterosexuality, in turn, is naturalised. Therefore, for lesbianism to occur it must come into existence through some disastrous situation.

Lesbianism is also seen as contributing to aggressive and disorderly behaviour among girls, and, once again, is not seen as being attributable to other factors such as poor conflict resolution skills: "What is more of a concern is that they are aggressive, operate in groups, stick together, and recruit younger students" (Harewood, 2010b, p. 14A). Due to gender socialisation, girls are not seen as or expected to be confrontational and the link between peer pressure and girls joining gangs, regardless of sexual orientation, is not made. Some girls are contesting the codes of femininity, and their gender transgression is being reduced to lesbianism. Therefore, gender and sexuality are conflated and are seen as one and the same.

In conclusion, this was a critical feminist perspective in theorising the relationship between gender, sexuality and lesbophobia in Caribbean culture. I have examined how lesbians are constructed through a heteropatriarchal gaze as "disruptive women" because they are perceived as violating dominant norms on gender and sexuality. Due to the overt homophobic violence directed towards gay men, it often goes unnoticed how lesbians are disciplined for contravening moralistic codes of heterosexual femininity, until sensationalist accounts appear in the media. Clearly, there needs to be a more nuanced or complex investigation of female sexuality that interrogates how different groups of women understand and experience their sexual lives.

Efforts launched to combat lesbophobia, and homophobia in general, must be multifaceted and account for how simultaneous oppressions related to gender and sexuality (along with race and class) produce a particular social reality for lesbians, who are positioned between two socially marginalised groups, women and LGBTQ people. Moreover, in forging strong alliances between feminist and LGBT groups in activism and organising, the links between heterosexism/homophobia and patriarchy/sexism, and actions to combat them, have be articulated as a major goal in the fight for social justice for all.

REFERENCES

Alexander, J. M. (2005). *Pedagogies of crossing: meditations on feminism, sexual politics, memory, and the sacred.* Duke University Press.

Alexander, M. J. (1991). Redrafting morality: The postcolonial state and the sexual offences bill of Trinidad and Tobago. In C. T. Mohanty, A. Russo, & L. Torres (Eds.), *Third world women and the politics of feminism* (pp. 133–152). Indiana University Press.

Alexander, M. J. (1997). Erotic autonomy as a politics of decolonization; an anatomy of feminist and state practice in The Bahamas tourist economy. In M. J. Alexander & C. Mohanty (Eds.), *Feminist genealogies, colonial legacies, democratic futures* (pp. 63–100). Routledge.

Atluri, T. (2001). *When the closet is a region: Homophobia, heterosexism and nationalism in the commonwealth Caribbean. Working paper no. 5.* Institute for Gender and Development Studies, Nita Barrow unit, the University of the West Indies.

Barriteau, E. (1998). Theorizing gender systems and the project of modernity in the twentieth century Caribbean. *Feminist Review, 59,* 186–210.

Barriteau, E. (2003). *Confronting power, theorizing gender: Interdisciplinary perspectives in the Caribbean.* University of the West Indies Press.

Barriteau, E. (2004). Constructing feminist knowledge in the commonwealth Caribbean in the era of globalization. In B. Bailey & E. Leo-Rhynie (Eds.), *Gender in the 21ˢᵗ century: Caribbean perspectives, visions and possibilities* (pp. 437–465). Ian Randle Publishers.

Barrow, C. (1996). *Family in the Caribbean.* Ian Randle Publishers.

Benkov, I. (1998). Yes, I am a swan: Reflections on families headed by lesbians and gay men. In G. Coll (Ed.), *Mothering against the odds: Diverse voices of contemporary mothers* (pp. 113–133). The Guilford Press.

Braidotti, R. (1994). *Nomadic subjects: Embodiment and sexual difference in contemporary feminist theory.* Columbia University Press.

Butler. (2007). *Gender trouble: Feminism and the subversion of identity.* Routledge.

Cabezas, A. L. (2004). Between love and money: Sex, tourism and citizenship in Cuba and The Dominican Republic. *Signs: Journal of Women in Culture and Society, 29*(4), 987–1016.

Capezza, N. M. (2007). Homophobia and sexism: The pros and cons to an integrative approach. *Integr Psych Behav, 41,* 248–253.

Carr, R. (2005). Homosexuality and HIV/AIDS stigma in Jamaica. In *Culture, health and sexuality* (pp. 1–13).

Chin, T. (1997). "Bullers and battymen" contesting homophobia in black popular culture and contemporary Caribbean literature. *Callaloo, 20*(1), 127–141.

Clarke, E. (1999). *My mother who fathered me*. The University of the West Indies Press.

Clarke, R. (1997). Combatting violence against women in Caribbean. In A. M. Baasiliero (Ed.), *Women against violence breaking silence: Reflecting on the experience in Latin America and the Caribbean*. The United Nations Development Fund for Women.

Comeau, D. (1999). Lesbian non-biological mothering: Negotiating an (un)familiar existence. *Journal of the Association for Research on Mothering Fall/Winter, 1*(2), 44–57.

Crawford, C. (2011, Fall/Winter). The Continuity of Global Crossings: African-Caribbean Women And Transnational Motherhood," for Mothering and Migration: (Trans) Nationalisms, Globalization, and Displacement. *Journal of Motherhood Initiative (JMI), 2*(2) 9–25.

Crichlow, W. E. A. (2004). History, (re)memory, testimony and biomythography: Charting a Buller Man's Trinidadian past. In E. Reddock (Ed.), *Interrogating Caribbean masculinities: Theoretical and empirical analyses* (pp. 185–222). University of the West Indies Press.

De Moya, A. E. (2004). Power games and totalitarian masculinity in The Dominican Republic. In E. Reddock (ed.), *interrogating Caribbean masculinities: Theoretical and empirical analyses,* (pp. 68-102). *University of the West Indies Press, 2004,* 68-102.

Du Long, J. (2005). "Lesbian activists in Jamaica tell horror stories" and "lesbian attacks less visible". In *We News—Lesbian and Transgender—Wednesday.* Womensenews.org/story/lesbian-transgender/050309/lesbian-activists-in-jamaica-tell-horror-stories. Retrieved June 10, 2012

Elwin, R. (1997). *Tongues on fire: Caribbean lesbian lives and stories*. Canadian Scholars' Press.

French, J., & Cave, M. D. (1995). Sexual choice as human right: Women loving women. In *Paper presented at the critical perspective on Human Rights in Caribbean*. Trinidad and Tobago.

Glave, T. (2008). *Our Caribbean: A gathering of lesbian and gay writing from the Antilles*. Duke University Press.

Harewood, C. (2010a). Girls gone wild, a concern. In *Nation News* (p. 5A). Barbados Sunday Sun.

Harewood, C. (2010b). Schoolgirls look to each other for love: Counsellor worried about the rising practice of lesbianism. In *Nation News* (pp. 14A–15A). Barbados, Sunday Sun.

Harewood, C. (2010c). "Nicole on the straight path" and "crusades against lifestyle". In *Barbados: Nation News* (pp. 12A–13A). Sunday Sun.

Harewood, C. (2010d). "Lesbianism not of god" (compilation of evangelists). In *Barbados: Nation newspaper* (p. 11A). Sunday Sun.

Human Rights Watch. (2016). *Jamaica: Investigate murder of alleged lesbians.* Retrieved June 10, 2012, from www.hrw.org/news/2006/07/26/jamaica-investigate-murder-alleged-lesbians.

Kempadoo, K. (2003). Theorizing Sexual Relations in the Caribbean: Prostitution and the Problem of the "Exotic". In E. Barriteau (Ed.), *Confronting Power, theorizing Gender: Interdisciplinary Perspectives from the Caribbean* (pp. 159–185). UWI Press.

King, R. (2008). More notes on the invisibility of Caribbean lesbians. In T. Glave (Ed.), *Our Caribbean: A gathering of gay and lesbian writing from the Antilles* (pp. 191–196). Duke University Press.

Kitzinger, C. (1994). Problematizing pleasure: Radical feminist deconstructions of sexuality and power. In H. L. Radtke & H. J. Stam (Eds.), *Power/gender: Social relations in theory and practice* (pp. 159–209). SAGE Publications Inc.

Lewis, L. (2003). *The culture of gender and sexuality in the Caribbean* (pp. 94–125). University Press of Florida.

Lorde, A. (1984). *Sister outsider essays and speeches.* Crossing Press.

Mason, G. (2002). *The spectacle of violence: Homophobia gender and knowledge.* Routledge.

Massiah, J. (2004). Feminist scholarship and society. In B. Bailey & E. Leo-Rhynie (Eds.), *Gender in the 21st century: Caribbean perspectives, visions and possibilities* (pp. 5–34). Ian Randle Publishers.

Mohammed, P. (1998). The Caribbean family revisited. In C. Barrow (Ed.), *Caribbean Portraits: Essays on gender ideologies and identities* (pp. 164–175). Ian Randle Publishers.

Mohammed, P. (2002). *Gendered realities: Essays in Caribbean feminist thought.* UWI Press.

Murray, D. A. B. (2009). Bajan Queens, nebulous scenes: Sexual diversity in Barbados. *Journal Caribbean Review of Gender Studies, Issue, 3,* 1–20.

Reddock, R. (1994). *Women, labour and politics in Trinidad and Tobago.* Zed Books.

Rich, A. (1995). *Of woman born: Motherhood as experience and institution.* Norton & Company, Inc..

Richardson, D. (1998). *Theorizing heterosexuality.* Open University Press.

Roberts, D., Reddock, R., Douglas, D., & Reid, S. (2012). *Sex, Power & Taboo: Gender and HIV in the Caribbean and Beyond.* Kingston: Ian Randle.

Robinson, T. (2003). Beyond the bill of rights: Sexing the citizen. In E. Barriteau (Ed.), *Confronting power, theorizing gender: Interdisciplinary perspectives from the Caribbean* (pp. 231–261). The University of the West Indies Press.

Rothman, B. K. (1989). *Recreating motherhood: Ideology and Technology in a Patriarchal Society.* Norton & Company.

Silvera, M. (1992). Man royals and sodomites: Some thoughts on the invisibility of afro-Caribbean lesbians. *Feminist Studies, 18*(3), 521–532.

Smith, R. T. (1996). *The matrifocal family: Power, pluralism and politics.* Routledge.

Tinsley, O. N. (2011). What is a Uma? Women performing gender and sexuality in Paramaribo, Suriname. In F. Smith (Ed.), *Sex and the citizen: interrogating the Caribbean* (pp. 241–250). University of Virginia Press.

Wekker, G. (1997). One Finger Does not Drink Okra Soup: Afro-Surinamese Women and Critical Agency. In M. Jacqui Alexander and C. Mohanty. (Eds.), *Feminist Genealogies, Colonial Legacies, Democratic Futures* (pp. 330–352). London: Routledge.

Wekker, G. (2006). *The politics of passion: Women's sexual culture in the Afro-Surinamese diaspora.* Columbia University Press.

Williams, L. (2000). *Homophobia and gay rights activism in Jamaica* (pp. 106–111). Small Axe.

Williams, C. L., & Stein, A. (2002). *Sexuality and Gender.* Blackwell Publishers.

Women for Women. (2010). *Violence against LBT Women in Jamaica.* Retrieved January 7, 2012, from http://ilga.org/ilga/en/article/miePUwr1UN

CHAPTER 6

Decolonising and Queering Caribbean Families

Abstract Given the paucity of research on same-sex unions and families in the Commonwealth Caribbean, in this chapter I will investigate the different ways gay men and lesbians are negotiating relationships within their birth families as well as how they are attempting to create their own families towards intimate and material security despite non-recognition from the state. First, I want to consider how sexual diversity and same-sex intimacy have always existed in the Caribbean, and elsewhere in the black Atlantic, through creolised representations of gender and sexuality. Second, I will examine how Caribbean families, in particular characteristics of working-class African-Caribbean families, have flexible socio-sexual unions and familial and kinship formations that deviate from a heterosexual Eurocentric nuclear family model which might aid in queering and decolonising dominant notions of family to be more inclusive of LGBTQ families in the region. Finally, drawing on the experiences of gay men and lesbians in Barbados, I want to problematise the same-sex unions and marriage equality in the region to uncover the different perspectives that Caribbean LGBTQ people have on this issue.

Keywords African-Caribbean families • Visiting unions and cohabitation • Extended family and kinship • Heteronormative sexual citizenship • Religion and nuclear familial ideology • Queer families,

© The Author(s), under exclusive license to Springer Nature Switzerland AG 2025
C. Crawford, *Gender, Sexual Citizenship and Epistemic Injustice in the Caribbean*, https://doi.org/10.1007/978-3-031-83493-6_6

affect and fictive kinship • Same-sex unions/partnerships • Marriage equality • Gay attitudes towards marriage

Sexual citizenship encompasses intimate life and family but heteronormativity in Caribbean societies makes it difficult for LGBTQ persons to create and sustain social partnerships and families of their own in safe and supportive social environments. Scholarship on Caribbean families, especially working-class African-Caribbean families, is vast but there has been little attention to how LGBTQ individuals establish sexual and affective bonds with others. Sociological and anthropological perspectives on the Caribbean have examined the family and household patterns of working-class African-Caribbean people, accounting for some of the notable non-nuclear family formations such as matrifocality and female-headed households and the primacy of the extended family and kinship networks under colonialism and thereafter (Herskovits & Herskovits, 1947; Clarke, 1999; Smith, 1996; Blake, 1961; Rodman, 1971; Senior, 1991; Barrow, 1996; Sutton & Makiesky-Barrow, 2001). Historically, the characteristics of working-class Caribbean families comprise adaptive cultural features developed under the weight of slavery and colonialism that include African retentions, favouring consanguinity, extended family units, female economic independence and mother–child bonds, which contrast with Eurocentric middle-class norms that value conjugality and the nuclear family unit with a male breadwinner (Barrow, 1999). Scholars have also examined transnational motherhood and families in looking at how women migrated to Canada and the United States to find employment to provide for themselves and their families across borders in the late twentieth century (Chamberlain, 2003; Crawford, 2012; 2018; Olwig Fog, 1996).

While Caribbean scholars have investigated the non-Western characteristics of African—Caribbean familial units and socio-sexual unions, and Caribbean feminists have critiqued patriarchy in families in exploiting women's productive and reproductive labour and sexuality for the of benefit men, most of these investigations have been from a heteronormative perspective that does not include same-sex couples and their families. The epistemic and structural inequalities born of heterosexism, homophobia and transphobia that were discussed in the previous chapters are magnified when discussing the obstacles that same-sex and queer couples face when they seek recognition for their unions and families in the Caribbean.

Sexual citizenship includes the freedom to engage in and form sexual intimacies and affective bonds that cover reproduction, family, civic unions and/or marriage. Cornell rightly argues that "if one is a homosexual, the right to engage in homosexual activity, has everything to do with family, marriage and procreation, even if the standard rights of heterosexual engagement have been denied to gay and lesbian couples, and even if gays and lesbians seek other forms of intimate association" (1992, p. 289). Given the paucity of research on same-sex unions and families in the Commonwealth Caribbean, in this chapter I take a preliminary look at the different ways LGBTQ persons are negotiating relationships within their birth families as well as how they are attempting to create their own families towards intimate and material security despite non-recognition from the state. First, I consider how sexual diversity and same-sex intimacy have always existed in the Caribbean, and elsewhere in the African diaspora, through creolised representations of gender and sexuality. Second, I examine how Caribbean families, in particular characteristics of working-class African-Caribbean families, have flexible socio-sexual unions and familial and kinship formations that deviate from a heterosexual Eurocentric nuclear family model which might aid in queering and decolonising dominant notions of family to be more inclusive of LGBTQ families in the region. Finally, drawing on the experiences of LGBTQ individuals in Barbados, I want to problematise same-sex unions and marriage equality in the region to uncover the different perspectives Caribbean LGBTQ people have on this issue.

Interrogating Caribbean Families: Adaptability and Diversity

While the non-recognition of same-sex partnerships is enforced in law and by the state in the Commonwealth Caribbean, there is a gap in collective understanding of how LGBTQ persons are, and have been, great contributors to their societies as well as how their affective capabilities, care work and reproduction are intermeshed in daily life. The lack of understanding and recognition of this reflects heteronormative structural inequalities that obscure and delegitimise queer knowledges and intimacies. The modern-day nuclear heterosexual family is based on a European middleclass -sexual division derived from the nascent phase of capitalist development and puritan heteropatriarchal ideology. I argue that

Caribbean scholarship outlining the unique characteristics of working-class African-Caribbean families has been impactful, whether intentional or not, in decolonising ideas about the universality of Western familial systems. Enslaved Africans resisted their subjugation by concealing and retaining some of their customs associated with gender, extended family and kinship arrangements despite European cultural dominance. This clash of cultures alongside black deprivation and colonial mimicry by black elites to get ahead resulted in creolised social relations and familial formations. Merle Hodge (2002) in "We Kind of Family" explains the different ways intimate and affective bonds are forged with others based on a variation of social arrangements and socio-sexual unions (albeit heterosexual). The author states that "a family is an organization of people that provides for its member's material needs (food, clothing, and shelter), and their emotional needs (approval, acceptance, solidarity and warmth), and socializes the young. There are different kinds of groupings that perform the functions of family" (Hodge, 2002, p. 477). Thus, the distinction between family and household is blurred in the Caribbean. Barrow adds that "the conviction that the family, that is the co-resident nuclear unit, is natural, universal and essential has faced it greatest test in the Caribbean" (Barrow, 1996, p. x). There is a variety of family and household types from nuclear families—either married or common law—with a man and woman sharing a common residence, to visiting relationships whereby unmarried couples are sexually involved but do not share a common residence, to single-parent families and extended families with three or more generations sharing a common residence (Barrow, 1996; Mohammed, 1998; Senior, 1991). This expansive definition of family accounts for the adaptive features of African-Caribbean families that are centred around collectivism to support single women with children and others in need (Barrow, 1996; Clarke, 1999). The recognition of common law unions and visiting unions among heterosexuals, and the high percentage of households headed by women in the Commonwealth Caribbean, is noteworthy in demonstrating that many families and households in the Caribbean are not always assembled along traditional conjugal and nuclear familial norms. Robinson (2017) notes that Caribbean countries have attempted to account for the varied heterosexual socio-sexual arrangements in the region that are bound by intimacy, affect and reproduction that are not established under the institution of marriage. She states that "Barbados in its Family Law Act 1981 was the first country to do so for what it described as "unions other than marriage." These were unions between a man and a woman living together continuously for at least five years (Robinson, 2017, p. 67). Other countries, such as Trinidad and Tobago, Guyana, Belize and Jamaica, followed suit in instituting cohabitation laws from the 1990s to early 2000s. Regardless of the familial and household type, women's economic contributions are integral to their families and households, challenging assumptions about the universal male breadwinner.

Black families in other parts of the African diaspora share similar flexible familial patterns that may create openings for LGBTQ individuals in establishing their own families. Chateauvert (2008) argues that "the prevalence of female-headed households among African Americans "queers" citizenship claims by challenging the Eurocentric and patriarchal nuclear family. In citizenship studies the idea of sexual or intimate citizenship offers the opportunity to re-frame analyses of African American families" (Chateauvert, 2008, p. 202). This is an important point in considering how the unconventionality of African-Caribbean families can be further decolonised and queered in opening spaces for gay men and lesbians to exercise their erotic autonomy to create and celebrate their socio-sexual unions and families regardless of a heteronormative status quo. In my earlier discussion of gender and sexual diversity in the Caribbean, I noted Lorde's (1982) account of female same-sex intimacy (zami) in Carriacou as well as Wekker's exploration of mati as "women who have sexual relations with other women, but who typically also will have had or still have relationships with men, simultaneously" (2009, p. 368). In addition to this, Tinsley (2011) locates the vibrancy of Afro-Surinamese female friendships and sexuality in creole culture, particularly in the oral tradition, by exploring rituals and Sunday gatherings through dance and songs (about flowers and gardens) in communal spaces or yards (dyari) in the post-emancipation period. Due to poverty and dis-enfranchisement, black people occupied public spaces in the urban centre where they could readily find work. Tinsley argues that there was not a strict division between public and private for black queer bodies. Instead of the "epistemology of the closet," there is what she calls "a Creole epistemology of gender and sexuality" (2011, p. 249).Tinsley goes on to argue that, due to race and class divisions, and out of spatial necessity, "nonheteronormative sexualities and complex gender (non) identities [were] not segregated from the everyday life of the dyari and the city, not shadowed by anxieties of being kept inside or outside" (2011, p. 249). In her seminal piece in "Man Royals and Sodomites: Some Thoughts on the Invisibility of Afro-Caribbean Lesbians," Silvera recounts her experiences of knowing women who loved

women while growing up in Jamaica. Despite the presence of heteropatriarchy and ideology and homophobia infused by Christian fundamental-
ism, her storied account of conversations between her mother and
grandmother inadvertently uncovers the complexity of female same-sex
love and desire in common spaces. Silvera grew up around "suspect" or
"different" women who dared to be themselves, strong, loud and sexual
while having intimate relationships either exclusively with women or with
both men and women amidst heterosexual normalcy. She rejected the
derogatory labels assigned to these women. "I remembered them. Not as
lesbians or sodomites or man royals, but as women that I liked. Women
whom I admired. Strong women, some colourful, some quiet" (Silvera,
2008, p. 351).

The racial and economic marginalisation under slavery and colonialism
that People of African Descent endured may have prevented black queer
sexualities from being totally separated from their communities despite
the disapproval of some individuals and the condemnation of churchgo-
ers. The black Atlantic queer experience is one of adaptation and not complete separation that is also witnessed in the United States. Growing up in
the Southern United States during segregation, bell Hooks (2015) notes
that collective survival against poverty and white supremacy may have
influenced the necessity of black queer people staying within their communities. She states that "sheer economic necessity and fierce white racism, as well as the joy of being there with the black folks known and loved,
compelled many gay blacks to live close to home and family" (Ibid.,
p. 121). hooks goes on to challenge the assumption that the black community is more homophobic than other racial-ethnic groups by discussing
the different ways that black queers negotiated their identities in the segregated South from some individuals leading closeted lives to others being
out in their families and communities. This example and others like it
reflect the existence, survival and resistance of black queer sexualities that
are located, albeit in contradictory ways, in racialised communities.
Moreover, the recognition of same-sex partnerships and their families is
critical in moving beyond solely focusing on queer intimacies as sex-based
acts, because this perspective obscures the affective and familial bonds that
they create as members of their communities. Furthermore, decolonising
and queering African-Caribbean families challenges affective-material
modalities of sexual citizenship that are tied to the productive, reproductive and consumptive patterns that privilege a middle-class heterosexual
nuclear family derived from a Eurocentric colonial capitalist project.

SAME-SEX RELATIONSHIPS AND FAMILY MATTERS

There is no universal way that LGBTQ individuals in the Commonwealth Caribbean go about establishing socio-sexual relationships. To problematise the relationship between sexual citizenship and intimate life for gay men and lesbians, I will revisit some of the experiences of LGBTQ persons in Barbados based on a study that was conducted between 2014 and 2015 as discussed in Chap. 3. I will, first, summarise some of the insights related to negotiating socio-sexual relations and attempting to create family under difficult circumstances and, second, I will analyse the accounts of gay men and lesbians who specifically discussed the need for the formal recognition of same-sex partnerships in Barbados. The experiences of LGBTQ persons in establishing relationships are not universal and attitudes about intimate citizenship are circumscribed by gender, class, age, race and educational background as well as social upbringing and environment. While some individuals know from childhood that they are gay and/or gender variant, it takes some time for them to articulate their feelings. This usually occurs during adolescence when emotional, mental and physical maturation takes place. The topic of teenage sexuality, and sexuality generally, must be demystified, to capture the social experience of Caribbean LGBTQ youth and educate them about safe and healthy sexual practices. LGBTQ youth tend to mask and suppress their feelings about their identities at home. But some individuals can carve out spaces with their friends and by engaging in extracurricular activities at school for validation and support.

It is important for Caribbean LGBTQ persons to be affirmed by their family and friends, and that includes having their significant others being recognised by those who are close to them. Some individuals may be reluctant to build long-lasting romantic relationships because they fear that their relationships will be sabotaged by those who disapprove of same-sex relationships. While heterosexuals may take it for granted that they can freely socialise and be affectionate with romantic partners in public, gay people do not have this privilege, so they are less likely to be demonstrative with their intimate partners in the public domain. Hanging out or "liming" in local parlance, is particularly challenging for effeminate gay men because they are usually ridiculed and punished for contravening hetero-masculine norms. There are limited spaces for queer folk to congregate, socialise and network in small communities so some individuals use online dating services while others organise house parties and social events by special invitation only. While private gatherings provide a certain degree of safety and anonymity for individuals, especially for those individuals with status and economic means, it nonetheless reinforces the limited choices that queer folk have in being out in public in their home countries. But some individuals can overcome this isolation by connecting to LGBTQ folks in the Caribbean diaspora in Britain, Canada and the United States via social media and by travelling abroad.

Creating a Family

Because of heteronormative constructions of family, not surprisingly, LGBTQ persons have a more flexible understanding of family that aligns with some of the characteristics of working-class Caribbean families in relation to extended family units, fictive kin, child fostering and polyamory. Working-class Caribbean families are eclectic in their features, which allows for multiple articulations of the function of family and the material and affective value that they provide in the care and maintenance of human beings. Thus, the heterogeneity of gay experience is seen through different socio-sexual unions and familial units that are class-based and culturally derived, which is a continuity of the unconventionality of creole Caribbean family systems.

Same-sex couples who cohabitate have strategically navigated the state's non-recognition of their unions and families by finding ways to celebrate their unions (commitment ceremonies) and by protecting their assets through joint ownership. While some individuals struggle to be in the company of their heterosexual friends and family members during special events like weddings, baby showers and anniversaries because such moments are not reciprocated for them, other individuals have created their own fictive kinship network that includes gay and straight people who they can celebrate special moments with. In the realm of social and biological reproduction, some queer persons may have children from previous heterosexual relationships, so they find ways to support and care for their children and integrate them into their new life whether they are their primary guardians or not. While gay people cannot legally adopt children in the Caribbean, some individuals help with the care of disadvantaged children in the family and community for a certain period, continuing the African-Caribbean practice of child-shifting or informal fostering. Despite the challenges that gay men and lesbians may have in creating their own families, there are opportunities for them to forge their own path in the realm of family and reproduction through a variety of ways including the use of reproductive technology.

Marriage Equality: I Do or I Don't

Currently, marriage in Barbados and other parts of the Commonwealth Caribbean is based on the recognition of the social partnership between a heterosexual man and woman by the state with or without religious authorisation. As noted before, heterosexual common law unions are social partnerships that are legally binding. Robinson surmises that "[…] still left in place in countries like Barbados is a dual justice family system which privileges heterosexual marriage and long-term heterosexual unions" (2017, p. 62). At this point in time, gay marriage or civil unions[1] are not legal in the Commonwealth Caribbean. The relationship between the church and state is strong in propagating a sexual citizenship that is heterosexual in theory and practice, which includes marriage being seen as a spiritual covenant between a man (husband) and woman (wife) who are brought together for the purpose of procreation. This situation is not unique to the Caribbean because marriage equality is still a work in progress throughout the world. Currently, 31 countries have legalised same-sex marriage with countries represented from both the global North and South (e.g. Canada 2003; South Africa 2006; Sweden 2009; Brazil 2011; France 2013; England and Wales 2014; Chile 2015; United States 2015; Costa Rica 2020 to name a few).[2] Discussions have started on same-sex civil unions or marriages in the region. Reid-Smith reports that "Barbados is one of the 20 countries included in a decision of the Inter-American Court of Human Rights. It ruled in 2017 that all its member states should legalize same-sex marriage" (2020, p. 6). Prime Minister Mia Mottley is supportive of same-sex unions and sees it as a step in the right direction in extending the country's commitment to human rights to LGBTQ persons. She states that "my government is prepared to recognize a form of civil unions for couples of the same gender so as to ensure that no human being in Barbados will be discriminated against, in exercise of civil rights that ought to be theirs" (Reid-Smith, 2020, p. 7). Structural inequality and hermeneutical injustice occur against sexual minorities when the legitimacy of marriage and family is exclusively tied to heterosexual conjugality

---

[1] Gay marriage refers to the recognition of same-sex relationships by the state and/or religious authority. Same-sex civil unions is used as a designation to recognise the rights of same-sex partnerships by the state that are equivalent to marriage but are not referred to as such because the latter term is retained for heterosexual partnerships.

[2] Marriage Equality Around the World. https://www.hrc.org/resources/marriage-equality-around-the-world

and reproduction. In addition to this, the non-recognition of gay marriages and civil unions may be economically burdensome to same-sex couples. Josephson states that "fundamental to my argument here is that intimate life and economic life are deeply intertwined, and that public policies that enforce an ideal version of sexual citizenship utilize economic incentives and disincentives to promote a particular version of desirable sexual citizenship" (p. 9). Thus, social injustice takes place against same-sex couples not only because of the non-recognition of their social partnerships by the state but also due to epistemic injustice that occurs in not valuing same-sex intimacies for their social, economic and affective contributions to their families, households and societies.

Given the different perspectives on marriage equality among LGBTQ persons in Barbados and in other parts of the region, there is no consensus on how this issue should be approached. There is hermeneutical friction among LGBTQ activists on the issue. On the one hand, legalising gay marriages or civil unions is important to some gay men and lesbians who want the same rights as heterosexual couples to legally establish long- term bonds with their partners that are bound in affective and material security that comes through the sharing of resources and property and in the raising of children. This position supports liberal feminist and queer perspectives that uphold the civil rights of all individuals to formally establish social partnerships under the law regardless of gender identity and sexual orientation (Thomas, 2017). But on the other hand, critical feminist and queer scholars are wary of this integrationist approach that adds women and sexual minorities to the institutions of marriage and family without reforming its structures that reproduce inequalities based on heteropatriarchal norms. Thomas argues that "far from what many sexual citizenship scholars and activists claim, marriage equality does not amount to queering sexual citizenship; instead, phallocentric citizens and institutions accommodated gays and lesbians by giving them one slice, albeit it a large slice, of heteronormative sexual citizenship" (2017, p. 570). Thus, some queer activists are reluctant to participate in the institution because they see it as de-radicalising the queer movement because *good gays* are supposed to appropriate the middle-class heteronormative rituals and lifestyle for their social partnerships to be legitimised. Besides these ongoing debates about marriage equality among feminist and queer activists, the issue of marriage equality is further complicated for individuals raised in religious families who may not have disclosed their sexual orientation to family members, let alone shared their desire to marry.

Overcoming Legal Hurdles

Joanne, a lesbian in her mid-20 s, shares her perspective on how same-sex couples are obstructed from forming social partnerships due to heteronormative values embedded in law and society. She states that "the law stipulates marriage is only for a man and a woman, and the expression for romantic love is only for a man and a woman." Since sexual citizenship is also a recognition and the legitimisation of the intimate and affective bonds that individuals establish with others, Joanne's account does not hold an idealistic view of same-sex marriage being legalised in Barbados any time soon. While Joanne takes a pragmatic position on the issue, Alex, a lesbian in her 30 s, hopes that she and her future partner will be able to celebrate their union in a special way despite state prohibition against gay marriage. Alex states that: "I'm hoping one day I could settle down with somebody that knows what they want. I mean well in Barbados it won't be like a marriage per se. But if I could have like exchanging of vows, and rings, and stuff, that would be beautiful commitment." Alex's account reflects on how intimate life for gay couples includes establishing long-term relationships with their partners and participating in rituals to celebrate and validate their commitment to each other. Sandy, a lesbian in her 20 s, also shares her perspective on the issue: "Obviously getting married and having kids as a lesbian couple here in Barbados I'm not able to do that. But I definitely will seek other means of making sure that I have my family and get what I need and what I know I deserve as human."

Some gay men and lesbians are indifferent to marriage equality because they feel that there is a lack of LGBTQ activism and public discourse on the matter in Barbados. Frank, a gay man, gives his insight:

> Even when it comes to gay marriages and … recently you've heard the Minister talk about not supporting gay marriages. And my thing is, who spoke about gay marriages… any gay people spoke about that? No, because it's something that the gay people really not that interested in. It's only because it's happening in the States and whatever happens overseas, comes here. It's not nothing that the gay community in Barbados is really interested in.

Based on Frank's account there seems to be hermeneutical confusion surrounding the support or lack of support for gay marriage in Barbados. Frank assumes that gay marriage is a non-issue in the Barbados and the region because he has not witnessed vigilant LGBTQ activism on it. He goes on to generalise about gay people's lack of interest in marriage equality and dismisses it as a foreign issue that has been picked up by opportunistic politicians who do not support it. Frank does not consider the challenges that LGBTQ groups may face in mobilising around this issue due to opposition by the state and religious gatekeepers or how some same-sex couples might be seeking to affirm their unions outside the institution of marriage. Gay couples look for support for their partnerships from Caribbean diasporic communities. For some couples, epistemic resistance entails migrating and settling permanently in another country like Canada or the United Sates to marry and start families, while others choose to get married abroad and then return home. Although their marriages are not recognised in Barbados, these couples nonetheless view the legalisation of their unions in another country as a symbolic gesture in validating their relationships. Paula married her partner in Canada and then returned to Barbados. She shares her thoughts on marriage equality and how she would like to live freely with her wife in her country:

> Yes. I would like for my legal status to be recognized in Barbados, but at the same time I want to live in an environment that I believe is accepting. Even if same sex marriage was to be legalized in Barbados, I still believe that the majority of persons would not be in favour of it. And I would like to be able to walk down the street and hold my partner's hand, my wife's hand. I would like to be able to kiss her in public if I so desire.

Paula understands that even if laws change to include gay marriage, it is harder to change people's prejudicial attitudes about homosexuality and gay couples. Thus, validating gay partnerships and families must be epistemically grounded in fairness by valuing their social experience of intimate life and family as a part of the whole.

While heterosexual couples can form intimate partnerships for emotional and affective reasons (or for love), there are material benefits that come along with marriage. Some gay couples who wish to legalise their unions are thinking about the long-term financial security of themselves, partners and children. Guy, a gay man in his 20 s who now lives in Canada, views gay marriage as of economic importance to same-sex couples: "Well first of all I would like to just see gay marriage legalized to the full extent where you get the same tax breaks, and you get the same insurance benefits." This gets back to Josephson's point that "that intimate life and economic life are deeply intertwined" (2016, p. 9). The economic deprivation that gay couples and their families encounter put them on the margins of sexual citizenship compared to heterosexual couples and families. Joanne, a lesbian, also feels that same-sex couples are economically disadvantaged when their partnerships are not legally recognised: "Gay marriage needs to be ok, because you have gay couples who've been together for years, but they do not have the same rights that other like common law relationships. If that person dies, we're screwed, especially if they're the breadwinner, we're screwed." Joanne's account reflects the unfair treatment that gay couples face compared to heterosexual couples when it comes to their partnerships and the allocation of resources to support them.

In conclusion, the expansion of sexual citizenship rights in the Caribbean for LGBTQ persons must include the recognition and protection of the rights of same-sex unions and families. Josephson makes a very important point that may help in working through the conundrum of marriage equality in the Caribbean. She states that "in my view the democratic state is and will continue to be engaged in the regulation of families and intimate life, but there are ways to make that regulation less inclined to enforce hierarchy and inequality, and to treat families and intimate life with greater fairness and justice" (2016, p. 8). Justice must be served in this area by the law and state to counter discrimination against sexual minorities regardless of the personal choices that they make about their partnerships and establishing a family. Progressive jurisprudence in decolonising and queering Caribbean families has already begun in recognising that the so-called traditional family derived from Victorian times is not fixed in time and space and that discriminatory laws cannot be upheld that abridge the right to privacy and family life of LGBTQ persons. I believe that the diversity and flexibility of Caribbean socio-sexual unions and families are dynamic enough to accommodate new imaginings of intimate life and family that include same-sex unions and their families in the twenty-first-century Caribbean. It is only a matter of time.

References

Barrow, C. (1996). *Family in the Caribbean: Themes and perspectives.* Ian Randle Publishers.

Blake, J. (1961). *Family structure in Jamaica.* The Free Press of Glencoe, Inc..

Chamberlain, M. (2003). Rethinking Caribbean families: Extending the links. *Community, Work & Family, 6*(1), 63–76.

Chateauvert, M. (2008). Framing Sexual Citizenship: Reconsidering the Discourse on African American Families. *The Journal of African American History, 93*(2), 198–222.

Clarke, E. (1999). *My mother who fathered me.* The University of the West Indies Press.

Cornell, D. L. (1992). Gender, sex and equivalent rights. In J. Butler & J. W. Scott (Eds.), *Feminists theorize the political* (pp. 280–296). Routledge.

Crawford, C. (2012). Who's your mama? Transnational motherhood and African-Caribbean women in the diaspora. In E. Barriteau (Ed.), *Love and power: Caribbean discourses on gender* (pp. 323–353). The University of the West Indies Press.

Crawford, C. (2018). Decolonizing reproductive labour: Caribbean women, migration and domestic work in the global economy. Special issue on Caribbean transmigration in the 21st century: contemporary re-imaginings and globalizing conditions, S. Gomes and M. Jokhan (Eds.). *The Global South, 12*(1), 33–55.

Herskovits, M. J., & Herskovits, F. S. (1947). *Trinidad village.* Alfred A. Knopf, Inc..

Hodge, M. (2002). We kind of family. In P. Mohammed (Ed.), *Gendered realities: Essays in Caribbean feminist thought* (pp. 474–485). The University of the West Indies Press.

Hooks, B. (2015). *Talking black: Thinking feminism, Thinking Black.* Routledge (reprint E-Book).

Josephson, J. J. (2016). *Rethinking sexual citizenship.* SUNY Press.

Lorde, A. (1982). *Zami a new spelling of my name: A biomythography.* Persephone Press.

Mohammed, P. (1998). The Caribbean family revisited. In C. Barrow (Ed.), *Caribbean Portraits: Essays on gender ideologies and identities* (pp. 164–175). Ian Randle Publishers.

Olwig Fog, K. (1996). The migration experience: Nevisian women at home and abroad. In C. Barrow (Ed.), *Family in the Caribbean: Themes and perspectives* (pp. 135–149). Ian Randle Publishers.

Reid-Smith, T. (2020). Barbados government proposes same-sex unions and hints it will make gay sex legal. *Gay Star News.* https://www.gaystarnews.com/article/barbados-government-proposes-civil-unions-and-hints-it-will-make-gay-sex-legal/

Robinson, T. (2017). Valuing care work. *Journal of Eastern Caribbean Studies Journal of Eastern Caribbean Studies, 42*(3), 59–79.

Rodman, H. (1971). *Lower class families: The culture of poverty in negro Trinidad.* Oxford University Press.

Senior, O. (1991). *Working miracles: Women's lives in the English-speaking Caribbean.* Indiana University Press.

Silvera, M. (2008). Man royals and sodomites: Some thoughts on the invisibility of Afro-Caribbean lesbians. In T. Glave (Ed.), *Our Caribbean: A gathering of lesbian and gay writing from the Antilles* (pp. 344–354). Duke University Press.

Smith, R. T. (1996). *The matrifocal family: Power, pluralism and politics.* New York, Routledge.

Sutton, C., & Makiesky-Barrow, S. (2001). Social inequality and sexual status in Barbados. In R. Reddock & C. Barrow (Eds.), *Caribbean sociology: Introductory Reading* (pp. 371–388). Markus Wiener Publishers.

Thomas, Q. J. (2017). Constructing queer theory in political science and public law: Sexual citizenship, Outspeech, and queer narrative. *New Political Science, 39*(4), 568–587.

Tinsley, O. N. (2011). What is a Uma? Women performing gender and sexuality in Paramaribo, Suriname. In F. Smith (Ed.), *Sex and the citizen: interrogating the Caribbean* (pp. 241–250). University of Virginia Press.

Wekker, G. (2009). Afro-surinamese women's sexual culture and the long shadows of the past. In C. Barrow, M. de Bruin, & R. Carr (Eds.), *Sexuality, social exclusion and human rights: Vulnerability in the context of HIV* (pp. 192–213). Ian Randle Publishers.

Index[1]

A

Alexander, M. Jacqui
  colonialism and policing black
    bodies, 25
  discriminatory laws against
    lesbians, 28
  erotic autonomy, 28
  heteropatriarchal recolonization, 27
  subjugated knowledges, 40
Anti-buggery laws, 4, 15, 25, 28, 32,
  49, 67, 69, 71–74, 83–85, 88,
  90, 92, 92n7, 94, 107
1533 Buggery Act, 25

B

Barbados' Domestic Violence
  (Protection Orders)
  Amendment Act, 51
Barbados Gays, Lesbians and All-
  Sexuals against Discrimination
  (B-GLAD), 65

Barriteau, Eudine
  gender systems, 27
Bulkan, Arif, 95

C

*Caleb Orozco v. The Attorney General of
  Belize*, Supreme Court of Belize
  Criminal Code, Section 53, 76
  decoloniality and religion, 77–80
  freedom of expression and equality
    under law, 83
  Protection of Fundamental Rights
    and Freedoms, 79
  re-interpretation of sexual
    citizenship, 85
  sexual identity, human dignity and
    the right to privacy, 80
  testimonial justice and overcoming
    silences, 76
  UNIBAM, 76
Canadian Charter cases, 74

[1] Note: Page numbers followed by 'n' refer to notes.

© The Author(s), under exclusive license to Springer Nature   145
Switzerland AG 2025
C. Crawford, *Gender, Sexual Citizenship and Epistemic Injustice in
the Caribbean*, https://doi.org/10.1007/978-3-031-83493-6

Caribbean feminism, 13
  cyber feminism, 66
  gender and nation-building, 27
  intersectional politics and
      feminism, 66
Citizenship, 6
  criticism, 7
  Marshall, T.H., 6
Colonial education
  missionaries, 24
  separate sphere/gender ideology, 24
Colonialism
  social control, 23

D
Decolonising and queering Caribbean
      families
  Afro-Surinamese female
      friendships, 133
  Caribbean families, adaptability and
      diversity, 131–134
  creole epistemology of gender and
      sexuality, 133
  extended family and kinship, 132
  female-headed households, 130
  marriage equality, 137–138
  same-sex civil unions, 137
  same-sex couples overcoming legal
      hurdles, 139–141
  same-sex relationships and family
      matters, 135–141
  visiting unions and the Family Law
      Act 1981, Barbados, 132

E
Elliott-Williams, Gabrielle
  decolonising Caribbean
      constitutionalism, 74
Enlightenment gender
  ideologies, 22, 91

Epistemic injustice, *see* Fricker,
      Miranda; Medina, Jose
Epistemic resistance, 63–64
  counter-publics, 65–66
  epistemic friction, 68
  intersectional politics and
      feminism, 66
  LGBTQ groups and activism, 65–68
  organising and capacity
      building, 67

F
Fricker, Miranda
  epistemic injustice, 41
  hermeneutical injustice, 42
  identity power, 41, 48
  testimonial injustice, 42

G
Gaskins, Joseph
  European voyages and
      'sodomites,' 25
Gender and citizenship
  UN Decade for Women, 26

H
Hermeneutical friction and political
      backlash, 69
  Sexual Offences Act, Barbados, 70
  Sexual Offences Bill, Jamaica, 69
Hermeneutical injustice, 42, 44, 50,
      52, 53, 57, 69, 80, 81,
      90, 98, 100
Hill Collins, Patricia, 41
  black feminism, 41
Hodge, Merle
  Caribbean families, 132
Homonationalism, 64
Homophobia, 16, 47, 55, 69–72

Homosexuality, ix, 10, 25, 32, 33, 47, 48, 54, 56, 70, 71, 77, 78, 82–84, 87, 89, 108, 110, 114, 117, 122, 140

I
International Resource Network, 67

J
Jamaica Forum for Lesbians, All-Sexuals and Gays (J-FLAG), 66
Jason Jones v. the Attorney General of Trinidad and Tobago, 85
    constitutionalism and sexual rights, 87
    right to privacy and same-sex intimacy & family, 90
    savings clause, 87
    social justice hermeneutics, 92
    state and church divided, 89
    unnatural acts and the homosexual outsider, 85

K
Kempadoo, Kamala
    colonialism and sex tourism, 23
    sexual praxis, 32
King, Rosamond
    cross-dressing in Carnival, 30–31
    lesbian invisbility, 46

L
Lazarus, Latoya
    Christian citizenship, ix, 28, 71, 78
Lesbianism, 106, 108, 114, 120, 121, 123
    female sexuality and the lesbian threat, 110–115

lesbophobia, 106–108
lesbophobia in media, 118–120
obscure lesbian subject, 108–110
queer lesbian identity, 114–115
violence against lesbians, 115–117
Lorde, Audre, 2
    Zami, 3

M
McEwan and Others v. Attorney General of Guyana, 93–94
    appeal to the Caribbean Court of Justice (CCJ), 99
    challenge to the cross-dressing ban, 95
    1893 vagrancy law, Section 153(I) (XLVII), 93
    hermeneutical virtue, 99
    Protection of the Fundamental Rights and Freedoms, 96
    sex and gender discrimination, 97
    structural hermeneutical inequality, 97
    transgender bodies crimalised as disorderly, 94
    transgender personhood, 100
Medina, Jose
    cognitive-affective responses, 44
    epistemic obstacles, 43
    epistemic resistance, 43
    hermeneutical resources, 43
    multiple publics, 43
    testimonial exchanges, 46
Murray, David
    sexual rights and human rights, 63

P
Pateman, Carol
    sexual contract, 8

Placide, Kenita, 68
  Eastern Caribbean Alliance for
    Diversity and Equality
    (ECADE), 68
  United and Strong, 68
Plantation societies, 21
  empire and erotic schemes, 22
  prostitution, 23
  sexual exploitation, 23

R
Racism
  anti-blackness, 22
  black sexuality, 22
Richardson, Diana
  conduct-based sexual rights, 31
  identity-based sexual rights, 31
Robinson, Tracy, 48, 66, 71, 97,
    109, 132
  gender and citizenship, 27
  gender neutrality in law, 50
  marriage, 137
  Robinson & Bulkan; cross-dressing
    ban, 98; decriminalisation, 95;
    strategic litigation, 73
  sexual citizenship, 62
  Sexual Offences laws, 28
  single parent family, 24
  UWI Rights Advocacy Project
    (U-RAP), 73

S
Savings clause, 75, 75n5, 87, 99
Sexual citizenship, 4, 5, 9, 21, 29, 64,
    66, 69, 72, 74, 75, 78, 82,
    85–87, 92, 134
  heteronormativity and redrafting
    morality, 28

institutionalization of
    heterosexuality, 10
  private/public dichotomy, 56
  sexual rights, 4
  Western hegemony, 20
Sexuality, 8
  commodification, 20
  Foucault, Michel, 10
  queer creole identities, 29
  queering, 30
  queer theory, 9n3
Sheller, Mimi, 4
  citizenship from below, 24
  counter-publics, 43
Silvera, Makeda, 133
Smith, Faith, 20
Society Against Sexual Orientation
    Discrimination (SASOD), 73
Strategic litigation, 73
  *Caleb Orozco v. The Attorney General
    of Belize*, Supreme Court of
    Belize, 75
  Caribbean Court of Justice
    (CCJ), 74
  decriminalising anti-buggery and
    vagrancy laws, 75–101
  *Jason Jones v. the Attorney General of
    Trinidad and Tobago*, 85
  *McEwan and Others v. Attorney
    General of Guyana*, 93–94

T
Testimonial injustice, 42–44, 46,
    47, 49, 50
Tinsley, Natasha, 133
Transatlantic slave trade
  chattel slavery, 21
Transgender, 29, 31, 64,
    93–95, 97

U

United Belize Advocacy Movement
(UNIBAM), 73
United Gays and Lesbians Against
AIDS in Barbados (UGLAAB), 65
UWI Rights Advocacy Project
(U-RAP), 73

V

Violence against LGBTQ, 29

W

Wekker, Gloria
mati, 30

The manufacturer's authorised representative in the EU is Springer
Nature Customer Service Centre GmbH, Europaplatz 3, 69115 Heidelberg,
Germany. If you have any concerns regarding our products, please
contact ProductSafety@springernature.com

Printed and bound by CPI Group (UK) Ltd, Croydon, CR0 4YY

27/04/2026

02097572-0003